T0258046

Deterioration, Structures and Analysis of Aluminium Alloys

Deterioration, Structures and Analysis of Aluminium Alloys

Edited by **Sally Renwick**

New York

Published by NY Research Press,
23 West, 55th Street, Suite 816,
New York, NY 10019, USA
www.nyresearchpress.com

Deterioration, Structures and Analysis of Aluminium Alloys
Edited by Sally Renwick

International Standard Book Number: 978-1-63238-116-3 (Hardback)

Printed in the United States of America.

Contents

Preface

There has been a breakthrough in the production of aluminium alloys. New procedures of welding, casting, forming and surface modification have emerged to advance structural integrity of aluminium alloys. The book covers major topics grouped under two sections namely, "Corrosion and Mechanical Damage on Aluminium Alloys" and "Micro-Nano Structures and Image Analysis". This book aims to serve the needs of a broad spectrum of professionals ranging from academic to industrial communities by providing latest information. It also serves the purpose of assisting technocrats, entrepreneurs and other individuals interested in the application and production of aluminium alloys.

Various studies have approached the subject by analyzing it with a single perspective, but the present book provides diverse methodologies and techniques to address this field. This book contains theories and applications needed for understanding the subject from different perspectives. The aim is to keep the readers informed about the progresses in the field; therefore, the contributions were carefully examined to compile novel researches by specialists from across the globe.

Indeed, the job of the editor is the most crucial and challenging in compiling all chapters into a single book. In the end, I would extend my sincere thanks to the chapter authors for their profound work. I am also thankful for the support provided by my family and colleagues during the compilation of this book.

Editor

Part 1

Corrosion and Mechanical Damage of Aluminium Alloys

1

Deformation Characteristics of Aluminium Composites for Structural Applications

Theodore E. Matikas[1] and Syed T. Hasan[2]
[1]Department of Materials Science and Engineering, University of Ioannina,
[2]Faculty of Arts, Computing, Engineering and Sciences, Sheffield Hallam University,
[1]Greece
[2]United Kingdom

1. Introduction

Silicon carbide (SiC) particulate-reinforced aluminium matrix composites (AMC) are attractive engineering materials for a variety of structural applications, due to their superior strength, stiffness, low cycle fatigue and corrosion fatigue behaviour, creep and wear resistance, compared to the aluminium monolithic alloys. An important feature of the microstructure in the Al/SiC composite system is the increased amount of thermal residual stresses, compared to unreinforced alloys, which are developed due to mismatch in thermal expansion coefficients of matrix and reinforcement phases. The introduction of the reinforcement plays a key role in both the mechanical and thermal ageing behaviour of the composite material. Micro-compositional changes which occur during the thermo-mechanical forming process of these materials can cause substantial changes in mechanical properties, such as ductility, fracture toughness and stress corrosion resistance.

The satisfactory performance of aluminium matrix composites depends critically on their integrity, the heart of which is the quality of the matrix/particle reinforcement interface. The nature of the interface depends in turn on the processing of the AMC component. At the micro-level, the development of local concentration gradients around the reinforcement can be very different to the nominal conditions. The latter is due to the aluminium alloy matrix attempt to deform during processing. This plays a crucial role in the micro-structural events of segregation and precipitation at the matrix-reinforcement interface.

The strength of particulate-reinforced composites also depends on the size of the particles, interparticle spacing, and the volume fraction of the reinforcement [1]. The microstructure and mechanical properties of these materials can be altered by thermo-mechanical treatment as well as by varying the reinforcement volume fraction. The strengthening of monolithic metallic material is carried out by alloying and supersaturating, to an extent, that on suitable heat treatment the excess alloying additions precipitates out (ageing). To study the deformation behaviour of precipitate hardened alloy or particulate reinforced metal matrix composites the interaction of dislocation with the reinforcing particles is much more dependent on the particle size, spacing and density than on the composition [2]. Furthermore, when a particle is introduced in a matrix, an additional barrier to the movement of dislocation is created and the dislocation must behave either by cutting through the particles or by taking a path around the obstacles [3].

At present, the relationship between the strength properties of metal matrix composites and the details of the thermo-mechanical forming processes is not well understood. The kinetics of precipitation in the solid state has been the subject of much attention. Early work on growth kinetics has been developed for the grain boundary case [4] and for intragranular precipitation [5]. These approaches have been integrated to produce a unified description of the inter- and intra-granular nucleation and growth mechanisms [6, 7]. More recently, successful attempts have been made to combine models of precipitate growth at interfaces with concurrently occurring segregation in aluminium alloys [8]. Studies of the relation between interfacial cohesive strength and structure have only recently become possible. This is due to of remarkable advances in physical examination techniques allowing direct viewing of interface structure and improved theoretical treatments of grain boundary structure.

The ability of the strengthening precipitates to support the matrix relies on the properties of the major alloying additions involved in the formation of these precipitates. The development of precipitates in Al-based alloys can be well characterised through heat treatment processing. Heat treatment affects the matrix properties and consequently the strain hardening of the composite. Furthermore, the distribution and concentration of these precipitates greatly affect the properties of the material where homogenous distribution of small precipitates provides the optimum results.

The role of the reinforcement is crucial in the microdeformation behaviour. The addition of SiC to aluminium alloy increases the strength and results in high internal stresses, in addition to the ones caused by the strengthening precipitates. Furthermore, the SiC reinforced particles are not affected by the heat treatment process. A great deal of attention has been recently devoted to understanding the strengthening mechanisms in metal matrix composites, which are distinguished by a large particulate volume fraction and relatively large diameter. Another important matter in understanding and modelling the strength of particulate MMCs is to consider the effect of particle shape, size and clustering [9-11], as well as the effects of clustering of reinforcement on the macroscopic behaviour and the effects of segregation to the SiC/Al interfaces [12]. Important role also play the effects of casting condition and subsequent swaging on the microstructure, clustering, and properties of Al/SiC composites [13].

Aluminium honeycomb sandwich panel constructions have been successfully applied as strength members of satellites and aircraft structures and also in passenger coaches of high-speed trains such as the TGV in France and the Shinkansen in Japan [14]. However, the cost of producing the all welded honeycomb structure has been a key factor for not using this technology on mass production rate. Recent developments in manufacturing methods have given rise to a range of commercially viable metallic foams, one being Alulight. In comparison to aluminium honeycomb core construction, metallic foams show isotropic properties and exhibit non linear mechanical deformation behaviour. The metallic foams have the potential to be used at elevated temperatures up to 200°C [15]. They also have superior impact energy absorption and improved strength and weight savings. However, the successful implementation of both aluminium honeycomb and metallic foam sandwich panels for aerospace and transportation applications is dependent upon an understanding of their mechanical properties including their resistance to fatigue crack growth and the resistance of aluminium alloys to environmentally induced cracking or stress corrosion cracking.

This chapter discusses first the relationship between the interfacial strength with the thermo-mechanical deformation process and the resulting macroscopic mechanical behaviour of particle-reinforced aluminium matrix composites. Micro-compositional changes which occur during the thermo-mechanical processing of these materials can cause substantial changes in mechanical properties such as ductility, fracture toughness, or stress corrosion resistance. A mico-mechanistic model will be presented for predicting the interfacial fracture strength in AMCs in the presence of magnesium segregation. Finally, the use of powerful nondestructive evaluation tools, such as infrared thermography, will be discussed to evaluate the state of stresses at the crack tip and to monitor fatigue crack growth in particle-reinforced aluminium alloy matrix composites.

In the second part of the chapter the structural integrity of Aluminium Honeycomb (HC) sandwich panels is compared with the new core material concept of aluminium foams. Aluminium Honeycomb sandwich panels are used to reduce weight whilst improving the compressive strength of the structure with the aerospace industry being one of the prime users of HC sandwich panels for structural applications. The cost of producing all welded HC structures has been the key factor for not using this technology on a mass production basis. An alternative to the aluminium honeycomb (HC) sandwich panels is the metallic foam sandwich panel, which has been gaining interest in the same field. These foams are anisotropic, exhibit non-linear mechanical behaviour, and they have the potential for use at temperatures up to 200°C. They have superior impact energy absorption, and improved strength and weight savings. The lower weight as compared to conventional solid wrought aluminium alloys will mean a reduction in fuel consumption thus providing economical savings.

This chapter attempts to investigate whether aluminium honeycomb sandwich panels, with their homogenous hexagonal core can be successfully replaced by metallic foam sandwich panels, which have an inhomogeneous core. A successful replacement would improve the confidence of manufacturers in the exploitation of this new material in replacing traditional materials. Current levels of understanding of cyclic stressing in metallic foam sandwich panels is limited and models of long term understanding of this aspect of failure are very important for both aerospace and automotive sectors. Burman et al [16] suggests that fundamental fatigue models and concepts proven to work for metals can be applied to metallic foam sandwich panels. A study by Shipsha et al [17] investigated experimentally both metallic foam and other cellular foams, using compact tension specimens. Shipsha's et al research is extremely interesting and implies that a sandwich panel should be considered whole and not two separate entities. Banhart and Brinkers has shown that it is very difficult to detect the features leading to fatigue failure in metallic foams due to the metallic foam being already full of micro cracks [18]. However, Olurin [19] investigation suggest that the fatigue crack growth mechanism of Alulight and Alporas foam is of sequential failure of cell faces ahead of crack tip. The main conclusion is that for a given ΔK, the fatigue crack propagation rate, da/dN decreases with increasing density and for a given stress intensity, the fatigue crack propagation rate increases when the mean stress is increased.

Current levels of understanding of cyclic stressing in aluminium foams is limited and models of long term understanding of this aspect of failure are important for both aerospace and automotive sectors. This is particularly important for low-density foam and honeycomb materials which despite thin ligament thickness, have good properties in compression. A method of analysis is proposed to predict life expectancy of aluminium honeycomb and metallic foam sandwich panels.

2. SiC-particulate reinforced aluminium matrix composites

2.1 Materials

Aluminium – silicon – magnesium alloys (A359) are important materials in many industrial applications, including aerospace and automotive applications. The alloys from the Al-Si-Mg system are the most widely used in the foundry industry thanks to their good castability and high strength to weight ratio. Materials based on A359 matrix reinforced with varying amounts of silicon carbide particles are discussed in this chapter.

Four types of material are used: 1) Ingot as received A359/40%SiC, with an average particle size of 19±1 micron, 2) Ingot as received A359/25%SiC, with an average particle size of 17±1 micron, 3) Hot rolled as received A359/31%SiC with an average particle size of 17±1 micron and 4) Cast alloy as received A359/30%SiC with particles of F400grit, with an average particle sizes of 17±1 micron. Table 1, contains the details of the chemical composition of the matrix alloy as well as the amount of silicon carbide particles in the metal matrix composites.

TYPES	Si	Mg	Mn	Cu	Fe	Zn	SiC
INGOT A359	9.5	0.5	0.1	0.2	0.2	0.1	40
INGOT A359	9.5	0.5	0.1	0.2	0.2	0.1	25
CAST A359	9.5	0.5	0.1	0.2	0.2	0.1	30
ROLLE D A359	9.5	0.5	0.1	0.2	0.2	0.1	31

Table 1. Types and composition of the material

The microstructure of such materials consists of a major phase, aluminium or silicon and the eutectic mixture of these two elements. In this system, each element plays a role in the material's overall behaviour. In particular, Si improves the fluidity of Al and also Si particles are hard and improve the wear resistance of Al. By adding Mg, Al – Si alloy become age hardenable through the precipitation of Mg_2Si particulates.

2.2 Heat treatment

Properties in particulate-reinforced aluminium matrix composites are primarily dictated by the uniformity of the second-phase dispersion in the matrix. The distribution is controlled by solidification and can be later modified during secondary processing. In particular, due to the addition of magnesium in the A359 alloy, the mechanical properties of this material can be greatly improved by heat treatment process. There are many different heat treatment sequences and each one can modify the microstructural behaviour as desired [20]. Precipitation heat treatments generally are low temperature, long-term processes. Temperatures range from 110°C to 195°C for 5 to 48 hours. The selection of the time temperature cycles for precipitation heat treatment should receive careful consideration. Larger precipitate particulates result from longer times and higher temperatures. On the other hand, the desired number of larger particles formed in the material in relation to their interparticle spacing is a crucial factor for optimising the strengthening behaviour of the composite. The objective is to select the heat treatment cycle that produces the most favourable precipitate size and distribution pattern. However, the cycle used for optimising one property, e.g. tensile strength, is usually different from the one required to optimise a different property, e.g. yield strength, corrosion resistance.

Heat treatment of composites though has an additional aspect to consider, the particles introduced in the matrix. These particles may alter the alloy's surface characteristics and increase the surface energies [21].

The heat treatments were performed in Carbolite RHF 1200 furnaces with thermocouples attached, ensuring constant temperature inside the furnace. There were two different heat treatments used in the experiments, T6 and modified-T6 (HT-1) [21, 22].

The T6 heat treatment consists of the following steps: solution heat treatment, quench and age hardening (Fig. 1).

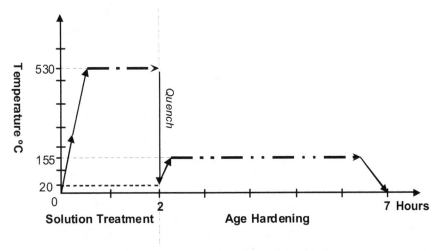

Fig. 1. T6 Heat treatment diagram showing the stages of the solution treatment for 2 hours and artificial ageing for 5h

In the solution heat treatment, the alloys have been heated to a temperature just below the initial melting point of the alloy for 2 hours at 530±5°C where all the solute atoms are allowed to dissolve to form a single phase solid solution. Magnesium is highly reactive with Silicon at this temperature and precipitation of Mg_2Si is expected to be formed. The alloys were then quenched to room temperature. In age hardening, the alloys were heated to an intermediate temperature of 155°C for 5 hours where nucleation and growth of the β' phase. The desired β phase Mg_2Si precipitated at that temperature and then cooled at room temperature conditions. The precipitate phase nucleates within the grains at grain boundaries and in areas close to the matrix-reinforcement interface, as uniformly dispersed particles. The holding time plays a key role in promoting precipitation and growth which results in higher mechanical deformation response of the composite.

The second heat treatment process was the modified-T6 (HT-1) heat treatment, where in the solution treatment the alloys have been heated to a temperature lower than the T6 heat treatment, at 450±5°C for 1 hour, and then quenched in water. Subsequently the alloys were heated to an intermediate temperature of 170±5°C for 24 hours in the age hardened stage and then cooled in air (Fig. 2).

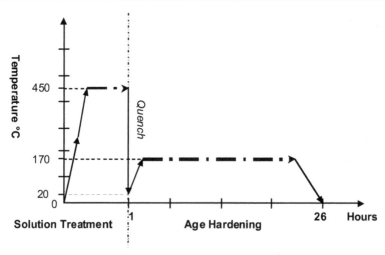

Fig. 2. Modified T6 (HT-1) showing stages of solution treatment for 1 hour and artificial ageing for 24h

In both heat treatments undesired formation of phases, like the Al_4C_3, is a possibility and selection of the solution treatment as well as the age hardening processes should be carefully considered. Temperature and time control, therefore, is extremely important during heat treatment. If the melt temperature of SiC/Al composite materials rises above a critical value, Al_4C_3 is formed increasing the viscosity of the molten material, which can result in severe loss of corrosion resistance and degradation of mechanical properties in the cast composite; excessive formation of Al_4C_3 makes the melt unsuitable for casting. In the A359/SiC composite high silicon percentage added in excess aids to the formation of some oxides (SiO_2) around the SiC reinforcement, something that retards the formation of Al_4C_3, since such oxides prevent the dissolution of SiC particles [22].

2.3 Metallographic examination
In order to analyse the microstructure, a series of sample preparation exercises were carried out, consisted of the cutting, mounting, grinding and polishing of the samples. The microstructures were investigated by SEM, EDAX, and XRD to determine the Al/SiC area percentage, size and count of particulates.

The microstructures of the examined MMCs in the as received condition have four distinct micro phases as clearly marked on the image micrograph, which are as follows: the aluminium matrix, the SiC particles, the eutectic region of aluminium and silicon and the Mg phase (Fig. 3). The distribution of SiC particles was found to be more or less uniform, however, instances of particle free zones and particle clustered zones were observed.

Matrix-reinforcement interfaces were identified by using high magnification Nano-SEM. In the as received hot rolled images the Al Matrix/SiC reinforcement interface is clearly identified (Fig. 4). These interfaces attain properties coming from both individual phases of constituents and facilitate the strengthening behaviour of the composite material.

In the modified T6 (HT-1) condition the microsturucture of the cast 30% SiC has the same phases as in the as received state, plus one rod-shape phase (Fig.5) along the matrix and at the matrix-reinforcement interface has been identified to be Mg_2Si precipitates in an early stage which are not fully grown. This evidence shows that β' phase has been formed with magnesium and silicon reacting together but β phases forming platelets of precipitates have not been formed in this HT-1 heat treatment, and this is probably due to the solution treatment temperature that did not allow enough reactivity time between the main alloying elements.

In the rolled 20% SiC the microstructure of HT-1 heat treatment shows an increase of the Silicon phase as shown in the image (Fig. 6). Silicon has been expanded during solidification and subsequent ageing. This formed round areas around the whole area of the composite. Comparing with the cast 30% SiC sample, in the rolled material the silicon phase is

Fig. 3. Microstructure of cast 30% SiC in the as received condition showing four distinct phases: Aluminium matrix, SiC particles, eutectic region of aluminium and silicon and Mg phase

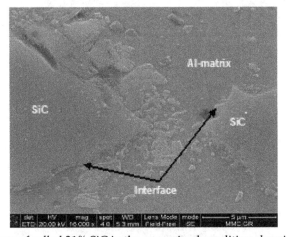

Fig. 4. Microstructure of rolled 31% SiC in the as received condition showing matrix-reinforcement interfaces

increased by 6%. This increase under the same heat treatment conditions is explained by the difference in the percentage of reinforcement in the material. Therefore, it becomes evident that the introduction of SiC reinforcement promotes zone kinetics and phase formation reactions during heat treatment process. The reinforcement, depending on its percentage in the matrix material, accelerates or restrains events such as precipitation and segregation. This is further supported by the fact that precipitation has not been observed in the HT-1 heat treated 20% SiC rolled material, where lower percentage of SiC reinforcement slowed-down the precipitation kinetics and β' phases could not be created in a similar manner as the 30% SiC cast sample.

In the T6 condition the microstructural results showed that in the rolled 31% SiC sample precipitates of Mg_2Si have been formed in the matrix in a platelet shape with a size of around 1-3 μm, as well as in areas close to the interface (Fig. 7). The higher solution temperature and lower age hardening holding time that exist in the T6 heat treatment process, promoted the forming of this type of precipitates which more densely populated

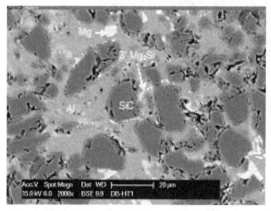

Fig. 5. Microstructure of cast 30% SiC in the HT-1 condition showing rod shape β' phases of Mg_2Si around the matrix and the interface of the reinforcement

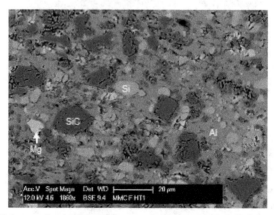

Fig. 6. Hot rolled HT-1 sample showing phases of Aluminium, SiC, Silicon, Mg

the interface region compared to the matrix. In the case of presence of a crack in the matrix, the precipitates act as strengthening aids promoting crack deflection at the interface resulting in an increase of the composite's fracture toughness [20, 23]. Furthermore, the precipitates formed in the matrix act as support to strengthening mechanisms of the reinforcement-matrix interface.

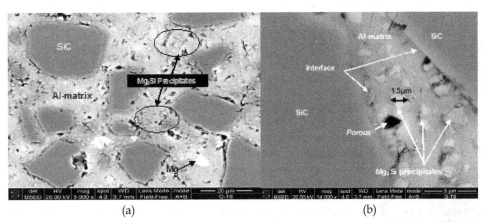

(a)   (b)

Fig. 7. (a) Hot rolled 31% SiC –T6 showing precipitate formed around the reinforcement. (b) Hot rolled 31% SiC – T6 showing Mg_2Si precipitates formed between the SiC reinforcement interface in a platelet shape of around 1-3 µm. A porous close to the interface has been identified in a similar size

The X-ray diffraction was carried out on the MMCs in the as received, as well as, in the heat treatment conditions, in samples with 20%, 30% and 31% of SiC particulates. Even though some peaks were superimposed, the results clearly showed the phases present in the microstructures. In particular, in the as received condition and in the heat treatment conditions the results showed existence of aluminium matrix material, eutectic silicon, SiC, Mg_2Si, SiO_2 phases as the distinct ones, and also $MgAl_2O_4$ and Al_2O_3 phases. $MgAl_2O_4$ and Al_2O_3 oxides give good cohesion between matrix and reinforcement when forming a continuous film at the interface. The presence of $MgAl_2O_4$ (spinel) shows that low percentage of magnesium reacted with SiO_2 at the surface of SiC and formed this layer in the interphacial region between the matrix and the reinforcement.

$$2SiO_2 + 2Al + Mg \rightarrow MgAl_2O_4 + 2Si \qquad (1)$$

The layers of $MgAl_2O_4$ protect the SiC particles from the liquid aluminium during production or remelting of the composites. This layer provides more than twice bonding strength compared to Al_4C_3. Furthermore, the layer of Al_2O_3 oxide is formed as a coating when SiO_2 is reacting with liquid aluminium.

$$3SiO_2 + 4Al \rightarrow 2Al_2O_3 + 3Si \qquad (2)$$

The same phases have been identified in the HT-1 modified condition. In the T6 condition XRD results showed one more phase present which is the spinel-type mixed oxide $MgFeAlO_4$ showing that Fe trace reacted with Mg and in the presence of aluminium and oxygen formed this oxide.

2.4 Micro-hardness testing

The three samples have been compared in relation to their microhardness performance based on the reinforcement percentage, the heat treatment conditions and the different manufacturing forming processes. Microhardness of the three composites has been measured in order to get the resistance of the material to indentation, under localized loading conditions. The microhardness test method, according to ASTM E-384, specifies a range of loads using a diamond indenter to make an indentation, which is measured and converted to a hardness value [21, 22].

Measuring the different phases in the micro-level it is quite challenging, as the SiC reinforcement of ≈17μm in size was not easy to measure, due to small indentation mark left when a small load on the carbide is applied. When introducing higher values of load, the indentation was not localized in the carbide but covered some of the matrix area too. The load was finally set to 50 grams in order to obtain valid measurements coming from different areas of the samples: SiC, aluminium matrix, and the overall composite i.e. areas superimposing matrix and reinforcement.

There are many factors influencing the microhardness of a composite material, including the reinforcement percentage, interparticle spacing and also particle size. Moreover, manufacturing forming processes influence material's microhardness behaviour in relation to the reinforcement percentages in the composites.

The cast sample in the as received condition has the highest MMC microhardness, where the rolled 20% SiC with lower percentage of reinforcement has the lowest values. By altering the microstructure with modified T6 (HT-1) heat treatment all values of the three samples show an increase between 20-45% from the initial state (Fig. 8). This shows the effect of the heat treatment in the micro-deformation of the matrix-reinforcement interface due to the presence of precipitates and other phases and oxide layers.

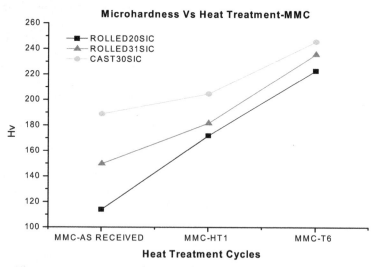

Fig. 8. Microhardness values Vs. Heat treatment cycles for the MMC areas

In the T6 condition it was observed the larger increase in microhardness values from the as received state, ranging from 20% to 90% depending on the reinforcement percentage and manufacturing process. In particular, in the rolled 20% SiC material the increase in microhardness values is in the order of 90%.

Furthermore, variability in microhardness values was observed when comparing cast and rolled materials with different percentage of SiC. However, this variability varied when samples processed at different heat treatment conditions were compared. Highest variability showed samples in the as received condition, whereas lowest variability showed samples in the T6 condition, with samples in the HT-1 condition in between. This can be explained by the fact that precipitates act as strengthening mechanisms and affect the micromechanical behaviour of the composite material.

In the absence of precipitates (in the as received condition), the volume percentage of SiC and the manufacturing processing play a significant role in micromechanical behaviour of the composite. As precipitates are formed due to heat treatment process they assume the main role in the micromechanical behaviour of the material. In the HT-1 heat treatment condition there is presence of β' precipitates which affect the micromechanical behaviour in a lesser degree than in the case of T6 heat treatment condition where fully grown β precipitates are formed. It becomes clear that after a critical stage, which if related to the formation of β precipitates in the composite the dominant strengthening mechanism is precipitation hardening.

While Figure 8 shows results in areas that include the interface region (where precipitates are concentrated) Figure 9, shows results on microhardness values in the aluminium matrix (where precipitates are dispersed). In Figure 9 there is similar variability for all three materials processing states, as received, HT-1, and T6. This implies that in the matrix material the percentage of the reinforcements, the manufacturing process, as well as the precipitation hardening, are strengthening mechanisms of equal importance.

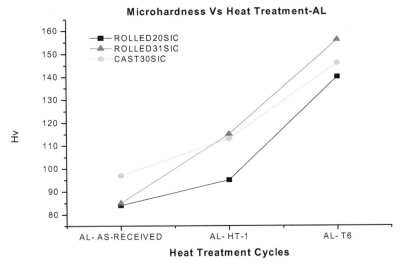

Fig. 9. Microhardness Vs. Heat treatment cycles for Aluminium areas

Figure 10 shows microhardness measurements obtained from areas around the matrix-reinforcement interface in a composite heat treated in the T6 condition. The microhardness values are higher in the close proximity of the interface. It is observed that cast material has higher values than the rolled material. In the case of rolled material, the microhardness raises as the percentage of reinforcement increases.

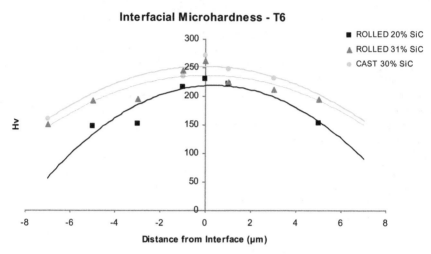

Fig. 10. Interfacial microhardness showing measurements obtained from areas close to the matrix- reinforcement interface in the T6 condition

2.5 Tensile testing

Aluminium – SiC particulate composite samples were tested in tension for two different volume fractions, 20% and 31%, in reinforcement [23]. The dog-bone coupons were tested according to ASTM E8-04 in the as received and, following two different heat treatments, modified T6 (HT-1) and T6 heat treatment conditions.. The mechanical properties of the composites are presented in Table 2.

Material	Condition	$\sigma_{0.2}$(MPa)	σ_{uts}(MPa)	ε(%)	E	$HV_{0.5}$
Rolled Al A359-SiC-20p	T1	146	157	1.5	100	114
	HT-1	147	190	4	102	172
	T6	326	360	2.1	112	223
Rolled Al A359-SiC-31p	T1	158	168	1	108	150
	HT-1	155	187	2	110	182
	T6	321	336	1.3	116	236

Table 2. The mechanical properties of Al/SiC Composites

The engineering stress/strain curves of the composite are shown in Figure 11. As can be clearly seen in Figure 11, the HT-1 heat treatment improved both the strength and strain to failure than the untreated composites for both volume fractions. Furthermore, the failure strain for this temper is considerably higher than for the T6 heat treatment; this may be attributed either to the nucleation of the β` precipitate phases which although not yet visible, may lead to the increase of the plastic deformation through crack deflection mechanisms and/or to annealing which acts competitively to the precipitation leading to the toughening of the composite. However, the T6 heat treatment exhibits the highest strength followed by the HT-1 and the as received state. Finally, as was expected, the "as received" composites behaviour in tension deteriorates with increasing filler concentration. The experiments showed that for the same range of conditions tested, the yield and the ultimate tensile strengths of the SiC/Al composites were mainly controlled by the percentage of reinforcement as well as by the intrinsic yield/tensile strengths of the matrix alloys. The addition of the SiC reinforcement created stress concentrations in the composite, and thus the aluminium alloy could not achieve its potential strength and ductility due to the induced embrittlement. Composites in the as-received condition failed in a brittle manner with increasing percentage of reinforcement. As a result, with increasing reinforcement content, the failure strain of the composites was reduced as shown in Figure 11. From the above postulations it is obvious that the phase that dominates the mechanical behaviour of the composite is the precipitation phase created by age hardening while the reinforcement phase plays a secondary role.

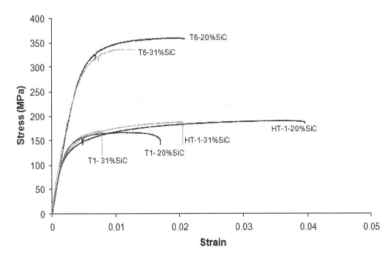

Fig. 11. Stress / Strain curves of Al/SiC Composites

The heat treatment affected the modulus of elasticity of the composites by altering the transition into plastic flow (see Table 2 and Fig. 12). Composites in the T6 condition strained elastically and then passed into a normal decreasing-slope plastic flow. Composites tested in the HT-1 condition exhibit a greater amount of strain than the as received and those heat treated in the T6 condition. The failure strain increasing from about 1.5% strain to about 4% but the greater influence was a sharper slope of the stress-strain curve at the inception of plastic flow.

This increase in elastic proportional strain limit and the steepening of the stress-strain curve were reflected by the higher yield and ultimate tensile strengths observed in the heat-treated composites. The increase in flow stress of composites with each heat-treatable matrix probably indicated the additive effects of dislocation interaction with both the alloy precipitates and the SiC reinforcement. The combination increased the strain in the matrix by increasing the number of dislocations and requiring higher flow stresses for deformation, resulting in the higher strengths observed. Ductility of SiC/Al composites, as measured by strain to failure, is again a complex interaction of parameters. However, the prime factors affecting these properties are the reinforcement content, heat treatment and precipitation hardening.

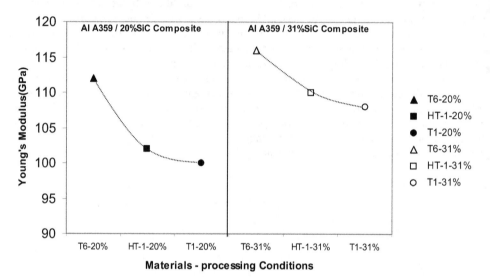

Fig. 12. Young's Modulus vs. Processing Conditions curves showing T6 treated composites having the highest modulus

2.6 Fracture toughness KIC testing

The plane strain fracture toughness test involves the loading to failure of fatigue pre-cracked, notched specimens in tension or in three-point bending. The calculation of a valid toughness value can only be determined after the test is completed, via examination of the load-displacement curve and measurement of the fatigue-crack length. The provisional fracture toughness value, K_Q, is first calculated from the following equation:

$$K_Q = \left(\frac{P_Q}{BW^{1/2}} \right) \cdot f \left(\frac{a}{W} \right)$$

(3)

where P_Q is the load corresponding to a defined increment of crack length, B is the specimen's thickness, W is the width of the specimen, and $f(\alpha/W)$ is a geometry dependent factor that relates the compliance of the specimen to the ratio of the crack length and width, expressed as follows:

$$f\left(\frac{a}{W}\right) = \frac{(2 + a/W)(0.86 + 4.64a/W - 13.32a^2/W^2 + 14.72a^3/W^3 - 5.6a^4/W^4)}{(1 - a/W)^{3/2}}$$

(4)

Only when specific validity criteria are satisfied, the provisional fracture toughness, K_Q, can be quoted as the valid plane strain fracture toughness, K_{IC}. The standard used for conducting this experiment, ASTM E399, imposes strict validity criteria to ensure that the plane strain conditions are satisfied during the test. These criteria include checks on the form and shape of the load versus displacement curve, requirements on specimen's size and crack geometry, and the 0.2% proof strength values at the test temperature. Essentially, these conditions are designed to ensure that the plastic zone size associated with the pre-crack is small enough so that plane strain conditions prevail, and that the linear elastic fracture mechanics approach is applicable.

Fracture toughness tests were conducted using a servo-hydraulic universal testing machine with data acquisition controller. The system was operated on load control for the fatigue pre-cracking stage, and on position control for the crack opening displacement (COD) testing. The fatigue test for pre-cracking was conducted at a frequency of 1 Hz, at a load ratio R = 0.25 and load range of 3.7 - 4.5 KN according to the materials' ultimate tensile strength. The COD was monitored by a clip gauge attached to the specimen with a testing rate set at 1 mm/min. Moreover, a thermal camera was set for thermographic monitoring of the crack opening displacement. Compact tension (CT) specimens were prepared for fracture toughness tests according to ASTM E399. The thickness B of the specimens was 9.2 mm for the MMC, and 5 mm for the unreinforced aluminium alloys.

Provisional K_Q values were calculated according to ASTM E399 standard for all specimens according to Equations (1) and (2), where Pq = Pmax. Load versus displacement curves for Al/SiCp composites and unreinforced aluminium alloys are shown in Fig. 13. Fracture toughness data for Al/SiCp and unreinforced aluminium alloys are summarised in Table 2.

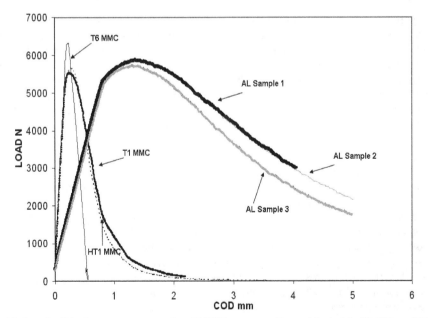

Fig. 13. Load – Displacement curves for Al/SiCp composites subjected to T1, T6 and HT-1 heat treatment conditions and three unreinforced aluminium alloy samples

Material	Heat Treatment	E (GPa)	$Rp_{0.2}$ (MPa)	B (mm)	a/W	α_{eff} (mm)	K_Q (MPa √m)	Valid	Reason
2000 series Al	AR	71	75	5.10	0.552	27.62	55,36	No	2**
2000 series Al	AR	71	78	5.13	0.555	26.76	56,00	No	2**
2000 series Al	AR	71	72	5.00	0.558	28.43	58,48	No	2**
A359/SIC/31p	T1	106	117	9.20	0.456	20.79	19,28	Yes	-
A359/SIC/31p	T6	108	120	9.21	0.462	20.12	22,05	Yes	-
A359/SIC/31p	HT1	116	157	9.20	0.467	21.33	20.75	Yes	-
A357/SIC/20p [16]	-	-	215	-	-	-	18.60	-	-
A359/SIC/10p [16]	-	-	300	-	-	-	17.40	-	-

**Validity criteria:
1 Excessive crack curvature
2. Thickness criteria not satisfied
3. Excessive plasticity
4. a/W out of range
5. Non-symmetrical crack front
6. In plane crack propagation

Table 2. Fracture toughness data for Al/SiCp and Al alloys and test validity

As is shown in Table 2, Al/SiC$_p$ composites exhibited lower provisional K_Q values than the reference unreinforced aluminium alloys. In addition, heat treatment processing, and especially T6 treated specimen, had the highest K_Q values compared to the other two heat treatment conditions. According to the load-displacement curves in Figure 3, composites clearly showed more brittle behaviour than the unreinforced aluminium alloys. T6 heat treated composites have the highest strength, but the lowest ductility compared to the other materials. Although these results provide some insight regarding the fracture behaviour of the materials examined, specific validity criteria have to be satisfied in order to obtain K_{IC} values.

2.7 Examination by infrared thermography

Nondestructive evaluation techniques are powerful tools for monitoring damage in composite materials [24]. Infrared thermography was used to monitor the plane crack propagation behaviour of particulate-reinforced AMCs [25, 26]. The deformation of solid materials is almost always accompanied by heat release. When the material becomes deformed or is damaged and fractured, a part of the energy necessary to initiate and propagate the damage is transformed in an irreversible way into heat [26]. The heat wave, generated by the thermo-mechanical coupling and the intrinsic dissipated energy during mechanical loading of the sample, is detected by the thermal camera. By using an adapted detector, thermography records the two dimensional "temperature" field as it results from the infrared radiation emitted by the object. The principal advantage of infrared thermography is its noncontact, non-destructive character.

A rectangular area on the specimen, located just in front of the initial pre-cracking region, was selected, as shown in Fig. 14a.

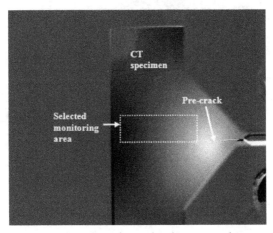

Fig. 14.a CT specimen showing the selected area for thermographic monitoring

The development of fracture was monitored in that area using infrared thermography. The mean temperature in this area versus time during crack growth was calculated using the recorded thermal imprint. As the specimen was stretched in tension, stresses were accumulating in the specimen, and the temperature increased as a function of time. When the accumulated energy became sufficient to propagate the crack, crack growth was observed, resulting in the stress relief. This corresponded to a peak in the temperature-time curve followed by a sudden decrease in temperature. As shown in Fig. 14b, 14c and 14d this behaviour was recurrent until the failure of the specimen. In these figures the thermographic monitoring of Aluminium 2xxx alloy, Al/SiCp T6 composite, and Al/SiCp HT1 composite samples is presented respectively. The different stages of crack growth for each material up to the final fracture of the specimen can be clearly observed. Just prior to fracture, the plasticity zone was clearly delineated on the specimen's surface as a heated region, which may be readily attributed to local plastic deformation. Furthermore, as seen in all figures, the crack was propagated in-plane throughout the experiment.

A comparison of the thermography graphs in Figs. 14b, 14c, and 14d leads to the conclusion that the aluminium alloy exhibited different crack propagation behaviour than the Al/SiCp composites. For the aluminium alloy, the temperature versus time curve in Fig. 5b showed extended plasticity behaviour before final fracture occurred. This behaviour was evidenced by the constant increase in temperature between the temperature picks at the 60th and 140th second (figure 14b). This behaviour may be attributed to the small specimen thickness. However, for the T6 heat treated composite material in Fig. 14c, fracture was more elastic as the multiple temperature peaks indicated a confinement of the plasticity zone. Also, plasticity was formed in a more balanced way regarding the overall fracture process. It was also observed that T6 heat treated composites exhibited fewer picks compared to the HT1 heat treated specimens (Fig. 14d). This was attributed to the presence of a stronger interface in the T6 material as the accumulation of precipitates near the interface, resulted in the improvement of the fracture toughness of the material.

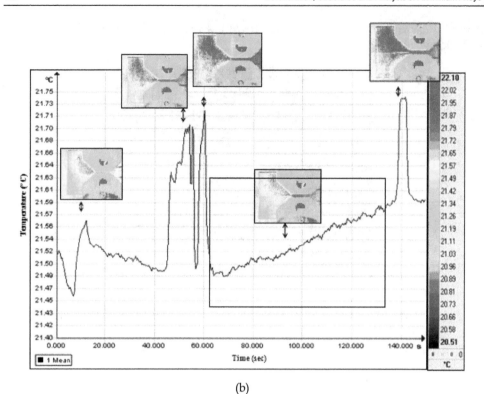

(b)

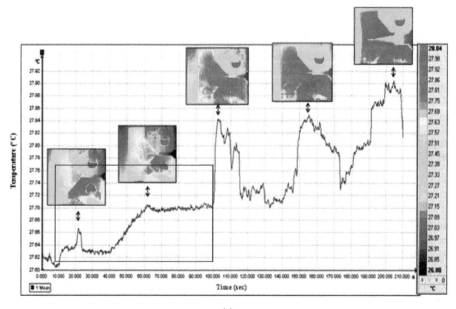

(c)

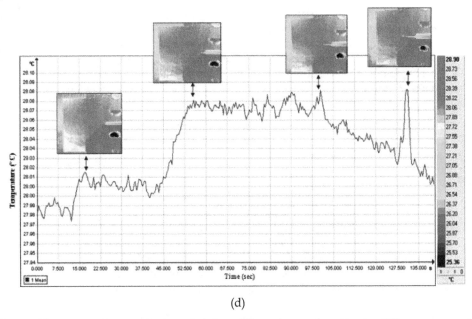

(d)

Fig. 14. Thermographic monitoring of various CT specimens showing the different stages of crack growth up to the specimen's final fracture: (b) Aluminium 2xxx, (c) Al/SiCp T6 composite, (d) Al/SiCp HT1 composite

2.8 A model for predicting interfacial strengthening behaviour of particulate reinforced AMCs

A model proposed by McMahon and Vitek [27] predicts the fracture resistance of a ductile material that fails by an intergranular mechanism. Based on this model, an effective work parameter can be developed to predict fracture strength of an interface at a segregated state using Griffith crack-type arguments. The Griffith's equation, which was derived for elastic body, is applied here because it is assumed that the yielding zone size ahead of the crack is small enough and the fracture is governed by the elastic stress field. The model further assumes that small changes in interfacial energy caused by segregation of impurities at the interface will result in a much larger change in the work of fracture. This is due to the fact that the work of fracture must be provided by a dislocation pile-up mechanism around the advancing crack-tip on the interface. This implies that additional work must be provided to deform the material at the crack-tip in addition to the work needed to overcome the interface energy and to replace it with two surfaces. The definition of interfacial fracture strength, σ_{int}, is then given by:

$$\sigma_{int} = \sqrt{\frac{100\varepsilon_p E_{int}}{\pi d}} \tag{5}$$

where,
E is Young's modulus, d is the particle thickness, since it is assumed that cracks of the order of the particle size are present when considering crack propagation through the interface

and the particulate, ε_p is the energy required to create two fracture surfaces = $2\varepsilon_s - \varepsilon_{gb}$ (= ε_0), with ε_s, the surface energy, and ε_{gb}, the grain boundary energy.

The 100 ε_p component allows for dislocation interaction and movement ahead of the crack-tip in ductile materials. This refers to the work required for a total separation of the lattice planes, which is equal to the area under the force-extension curve.

From equation (6) ε_p can be estimated if Kint (Interface fracture toughness) and Eint (Interface Young's Modulus) are known.

$$\frac{K_{int}^2}{100E_{int}} = \varepsilon_p \left(1 - \frac{ZRT\ln(1-c+Bc)}{\varepsilon_p} \right)^n \tag{6}$$

Where,

Z, describes the density of interface sites which are disordered enough to act as segregation sites (= D ρ_s), with D the thickness of the interface region, and ρ_s the density of the interface region (D=300 nm) (ρ= 2.6889 g/cm^3 for Aluminium and 3.22 g/cm^3 for SiC),

R, is the gas constant (= 8.314472(15) J•K-1•mol-1),

T, is the absolute temperature (= 803.15 K for T6, = 723.15 K for HT1),

c, is the segregate concentration needed to cause embrittlement (= 0.1 for pure aluminium),

B, describes the modification of the boundary energy by impurities using the Zuchovitsky equations,

n, is the work hardening exponent (= 10 for FCC aluminium).

In hard particle reinforced metal matrix composites the stress transfer from the matrix to the particles is mainly controlled by the misfit of the elastic constants between the two phases [28]. To measure the stress transfer to the particle, in an homogeneous material subjected to tensile loading, the stress carrying capability of the particle is defined as the ratio of the normal stress σ_N to the particle in the loading direction to the macroscopic tensile stress, σ_T, i.e. the ratio L = σ_N / σ_T. By using Eshelby's theory, the stress carrying capability of a spherical inhomogeneity can be written as [29]:

$$L = \frac{9x(2+3x)}{(1+2x)(8+7x)} \tag{7}$$

where, x = E_i / E_m, and E_i and E_m are Young's moduli for inhomogeneity and matrix, respectively.

Furthermore, the shear lag model, originally developed by Cox [30] modified by Llorca [31], can be used to estimate the stress carrying capability of a particulate, assuming that the volume fraction of reinforcement is small and the average stress in the matrix is approximately equal to the applied stress:

$$L = 1 + \frac{a}{\sqrt{3}} \tag{8}$$

where $a = \bar{h}/2\bar{r}$ is the aspect ratio of the reinforcement, with $\bar{h}$ and $\bar{r}$ the average length and the average diameter of the particle.

A model has been proposed to estimate the effects of particle volume fraction on fracture toughness in SiC particle-reinforced aluminium alloy matrix composites. This model assumes that SiC particles are uniformly distributed in the matrix and that the pattern of particle distribution is similar to FCC structure in metals. The fracture toughness of the composite can then be written as [32]:

$$K_{IC} = \frac{K_p}{L_p}V'_m + \frac{2K_{int}}{L_p + L_m}(V_m - V'_m) + \frac{K_m}{L_m}2V_m + K_m(1 - 3V_m)$$

(9)

where K_{IC}, K_p = 3 MPa m-1/2, K_m = 35 MPa m-1/2, and K_{int} is the fracture toughness of the composite, SiC particulates, A359 aluminium alloy matrix, and interface, respectively. L_p and L_m are the stress carrying capabilities of a particulate and the matrix, respectively. On average, for SiC particles and aluminium alloy matrix, $L_p \sim L_m \sim 2$. The value of $L_m = 1$ is applicable for clean surfaces. However, due to processing conditions and the physical interaction at the matrix/reinforcement interface the interface contains partially contaminated surfaces, therefore $L_m = 2$ since it cannot be considered as a "clean surface". V_m and $(V_m - V'_m)$ are the area fractions for particle cracking and interface failure, respectively. These fractions though are not accurately known. However Wang and Zhang [33] found that the ratio of particle cracking over interface failure $V_m/(V_m - V'_m)$ was about 0.13 (= 1.4%/10.7 %) in a SiC particle-reinforced aluminium alloy composite.
Young's modulus of matrix has been obtained for A359 aluminium matrix. The particles E_p, matrix E_m, and interface E_i shown in equation

$$E_C = E_p v_f^{2/3} + E_m\left(1 - V'^{2/3}_f\right) + E_i\left(V'^{2/3}_f - V_f^{2/3}\right)$$

(10)

Due to the fact that the difference $\left(V'_f - V_f\right)$ is very small, a good approximation is to consider that the Young's modulus of the interface region is close to that of the matrix; $E_i \cong E_m$ [32].
The parameter B describes the modification of the boundary energy by impurities using the Zuchovitsky equations [34, 35], given by:

$$B = e^{\left(\frac{\varepsilon_1 - \varepsilon_2}{RT}\right)} \cong e^{\left(\frac{0.75\varepsilon_F}{RT}\right)}$$

(11)

where $\varepsilon_2 - \varepsilon_1$ is the difference between the formation energy in the impurity in the bulk and the interface region. It is assumed that the values of the surface energy and the impurity formation energy in the bulk are close, and therefore the numerator in the exponential term depends on the impurity formation energy in the interface region, which is assumed to be 0.75 ε_f, where ε_f is the formation energy of the impurity in the bulk.
Using Faulkner's approach [36], to the derivation of impurity formation energy,

$$\varepsilon_f = \varepsilon_s + \varepsilon_e$$

(12)

where, ε_s is the surface energy required forming the impurity atom and ε_e is the elastic energy involved with inserting an impurity atom into a matrix lattice site. This is given by:

$$\varepsilon_f = \frac{0.5\varepsilon_S}{1.94} + \frac{8\pi G}{3e}a_m\left(a_i - a_m\right)^2 eV$$

(13)

where,
ε_S is the surface energy (1.02 J m-2)

e is the electronic charge (1.60217646 *10^19 Coulomb)

a_i is the impurity atomic radius (0.118 nm for Si)

a_m is the matrix atomic radius (0.143 nm for aluminium)

G is the shear modulus (26 GPa for aluminium)

By performing the calculations the impurity formation energy, ε_f, for A359 aluminium alloy (Al-Si-Mg) can be determined and then substituted in equation (11) to calculate B.

The micro-mechanics model described above is based on thermodynamics principles and is used to determine the fracture strength of the interface at a segregated state in aluminium matrix composites. This model uses energy considerations to express the fracture toughness of the interface in terms of interfacial critical strain energy release rate and elastic modulus. The interfacial fracture toughness is further expressed as a function of the macroscopic fracture toughness and mechanical properties of the composite, using a toughening mechanism model based on stress transfer mechanism. Mechanical testing is also performed to obtain macroscopic data, such as the fracture strength, elastic modulus and fracture toughness of the composite, which are used as input to the model. Based on the experimental data and the analysis, the interfacial strength is determined for SiC particle-reinforced aluminium matrix composites subjected to different heat treatment processing conditions and the results are shown in table 2. It is observed that K_{int} values are close to the K_{1c} values of the composites. Furthermore, σ_{int} values found to be dependent on the heat treatment processing with T6 heat treatment composite obtain the highest interfacial fracture strength.

2.9 Fatigue testing and crack growth behaviour

Tension-tension fatigue tests were conducted using a hydraulic testing machine. The system was operated under load control, applying a harmonic tensile stress with constant amplitude. By specifying the maximum and the minimum stress levels, the other stress parameters could be easily determined. These were the stress range, σ_r, stress amplitude, σ_a, mean stress, σ_m, and fatigue stress ratio, R ($=\sigma_{min}/\sigma_{max}$). Throughout this study, all fatigue tests were carried out at a frequency of 5 Hz and at a stress ratio R = 0.1. Different stress levels between the ultimate tensile strength (UTS) and the fatigue limit were selected, resulting in so-called Wöhler or S-N curves. Tests exceeding 10^6 cycles without specimen failure were terminated. Specimens that failed in or close to the grips were discarded. The geometry of the samples was the same as those used for the tensile characterisation, i.e. rectangular strips of 12.5mm width, and 1.55mm thickness.

The normalised "S-N" curves of the fatigue response of the Al/SiC composites is shown in Fig. 15. The stress was normalised over the UTS of each material and plotted against the number of cycles to failure. As can be observed, whereas in the untreated T1 condition the composite retains at least 85% of its strength as fatigue strength, the corresponding value for the T6 heat treatment is falling to the 70% of UTS. The HT1 heat treatment is exhibiting an intermediate behaviour, with its fatigue strength falling to 75% of the corresponding UTS value. It can be concluded that aggressive heat treatment reduces the damage tolerance of the composites.

A direct comparison of the fatigue performance of the composite with the corresponding quasi static performance in tension reveals that the T6 heat treatment improved the strength of the composite. This can be attributed to a dominant mechanism related to the changes in the microstructure of the composite. This mechanism relates to the precipitations appearing in the microstructure of the composite at the vicinity of the interphase area, which results to the composite hardening. The creation of the interphase together with the improved stress transfer may be regarded as the main contributing parameters to the improved mechanical properties of the particulate reinforced composite. The improved static strength is followed by a less spectacular performance in fatigue, with the fatigue limit of the material falling to the 70% of the UTS.

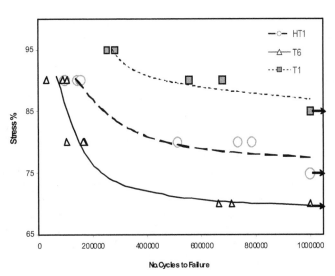

Fig. 15. S-N Curve of Al/SiC 20% Composite

2.9.1 Crack growth rate vs. range of stress intensity

To study the crack growth rate (da/dN) vs. stress intensity range (ΔK) data for aluminium SiC$_p$ composites and aluminium 2xxx series specimens, the materials were subjected to cyclic loading. Fatigue crack growth tests were conducted according to the ASTM E647 standard using a servo hydraulic testing machine. The tests were conducted under load control. Compact tension (CT) specimens were prepared for the fatigue crack growth experiments. The fatigue tests for the monolithic aluminium specimens were performed at a standard frequency of 5 Hz. However, a lower frequency of 1 Hz was selected for the composite specimens in order to minimize the effect of sudden failure due to the brittle nature of these materials. The experiments were performed at a load ratio R = 0.25 and maximum load ranges of 3.7 - 4.5 KN, keeping the maximum stress at about 70% of the material's ultimate tensile strength.

The technique used for determining the crack growth rate versus stress intensity range during the cyclic loading tests was based on non-contact monitoring of the crack propagation by lock-in thermography. This new technique deals with mapping the crack growth nondestructively. Lock–in thermography is based on remote full field monitoring of thermal waves generated inside the specimen by cyclic loading that caused an oscillating temperature field in the stationary regime. Lock–in refers to the necessity for monitoring the exact time dependence between the output signal (thermal wave) and the reference input signal (fatigue cycle). This is done with a lock–in amplifier so that both phase and magnitude images become available.

The detection system included an infrared camera. The camera was connected with the lock-in amplifier, which was then connected to the main servo hydraulic controller. Therefore, synchronization of the frequency through the lock-in amplifier and the mechanical testing machine could be achieved and lock-in images and data capture during the fatigue testing were enabled.

The camera was firstly set at a distance close to the specimen, in order to have the best possible image capture. Then, the fatigue pre-cracking started while synchronizing, at the same frequency, fatigue cycles and infrared camera through the lock-in amplifier.

In order to determine the crack growth rate and calculate the stress intensity factor using thermographic mapping of the material undergoing fatigue a simple procedure was used:

a. The distribution of temperature and stresses at the surface of the specimen was monitored during the test. Therefore, thermal images were obtained as a function of time and saved in the form of a movie.

b. The stresses were evaluated in a post-processing mode, along a series of equally spaced reference lines of the same length, set in front of the crack-starting notch. The idea was that the stress monitored at the location of a line versus time (or fatigue cycles) would exhibit an increase while the crack approaches the line, then attain a maximum when the crack tip was on the line. Due to the fact that the crack growth path could not be predicted and was not expected to follow a straight line in front of the notch, the stresses were monitored along a series of lines of a certain length, instead of a series of equally spaced points in front of the notch. The exact path of the crack could be easily determined by looking at the stress maxima along each of these reference lines.

Four lines of the same length, equally spaced at a distance of 1 mm, were set on the thermal images of the CT specimen at a distance in front of the specimen's notch.

In Figure 16, the crack growth rate for the heat treated composite specimens and the reference aluminium alloy samples are plotted on a logarithmic scale as a function of the stress intensity range. The results showed that the heat treatment processing influences crack growth behaviour of the composite materials. Specimens subjected to T6 heat treatment condition exhibited the highest crack growth rate vs. stress intensity range slope compared to the other composite systems. Moreover, the crack growth rate vs. stress intensity range line of specimens subjected to T6 heat treatment was shifted towards higher ΔK values compared to that from specimens subjected the other two heat treatment conditions. This implies that in order to attain the same crack growth rate, higher stress intensity factor is required for specimens subjected to T6 condition compared to those subjected to T1 and HT-1 conditions. The need for higher stresses for a crack to propagate reveals the material's microstructural strength, where micro-mechanisms such as precipitation hardening promote high stress concentrations at the crack tip, resulting in the toughening of the crack path. The above postulations agree with previews results, where the T6 heat treated composites showed superior strength but the lowest ductility compared to T1 or HT-1 heat treated specimens. Results, shown in Figure 16, indicate that at intermediate values of crack growth rate (10^{-2} to 10^{-5} mm/cycle) the Al/SiCp composites have fracture properties comparable to those of the unreinforced matrix alloys. It is obvious that in these composites crack propagation rate seems more balanced and takes more time than the aluminium alloy obtaining crack growth rate values from around 10^{-1} to 10^{-4} mm/cycle.

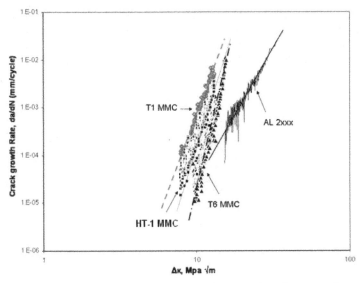

Fig. 16. da/dN vs. ΔK plots of Al/SiCp composite and monolithic aluminum 2xxx specimens

2.9.2 Estimation of da/dN vs. ΔK using thermography and compliance methods

Using the procedure described above based on thermographic mapping, the local stress versus time was measured for the T6 heat treated Al/SiCp along each of the four reference lines placed in front of the CT specimen's notch. The maximum value of stress versus the number of fatigue cycles was then plotted for those four lines (Figure 17). As expected, Figure 17 shows that the local stress, monitored at the location of each line, increases as the crack is approaching that line, then attains a maximum when the crack tip is crossing the line. Finally, after the crack has crossed the line, the local stress measured at the location of the line decreases. This is also expected, since the stress values shown in Figure 17 are stress maxima from all the locations along the particular line. At the exact position on a line where the crack has just crossed, the local stress is null as expected.

From the stress maxima versus fatigue cycles curves, for each reference line, shown in Figure 17, the crack lengths versus the number of fatigue cycles were determined for A359/SiCp composites in all three different thermal treatment conditions: T1, T6, and HT-1 (Fig. 18). As it is shown in Figure 18, the crack growth rate was found to be quite linear for all heat treatments. Also, there is a small change in the linear slope for the HT-1 heat treated sample, showing increased ductility, which indicates that more time (i.e. cycles) is needed for the crack to grow in this case. For the T6 heat treatment, the results depict a brittle behaviour, as the crack starts to grow earlier than in the other two cases, supporting evidence of brittle fracture.

The stress intensity range was further calculated by the data shown in Figure 17. ΔK values have been estimated from the stress maxima versus fatigue cycles curves for each reference line, shown in Figure 17. Each of the four lines provides a stress intensity range and a da/dN value. The data obtained using lock-in thermography, shown in Figures 18 and 19, were correlated with crack growth rate values obtained by the conventional compliance method and calculations based on the Paris law. Furthermore, the da/dN vs. ΔK curves steaming from the compliance method were plotted in the same graph, for comparison purposes, with those obtained using lock-in thermography (Fig. 20). It can be seen in Figure 20 that the two different methods are in agreement, demonstrating that lock-in thermography is a credible nondestructive method for noncontact evaluation of the fracture behaviour of materials.

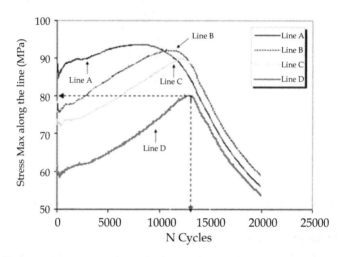

Fig. 17. T6 al/SiCp stress maxima along the four reference lines vs. number of fatigue cycles

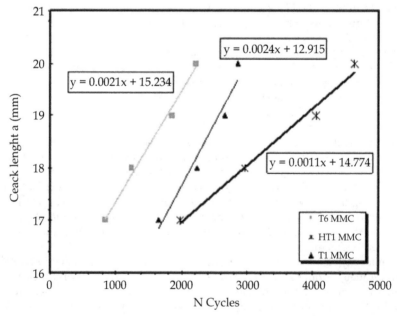

Fig. 18. Crack length vs. cycles obtained from lock-in thermography data for A359/SiCp composites subjected to three different heat treatment conditions

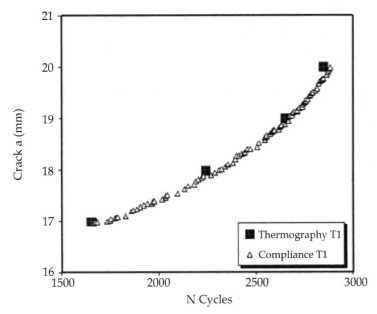

Fig. 19. Crack growth rate determined by compliance vs. thermography for A359/SiCp composite

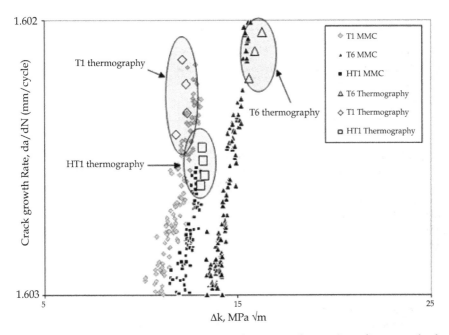

Fig. 20. da/dN vs. ΔK for Al/SiCp specimens - Thermography vs. Compliance method

3. Aluminium honeycomb sandwich panels and metallic foam

3.1 Material and experimental procedure

The compact tension specimens were manufactured from rectangular plates conforming to BS 7448-1: 1991 as shown in Fig. 21.

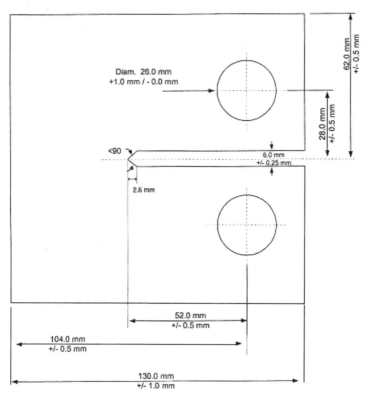

Fig. 21. Compact tension specimen

The fatigue tests were performed on a standard Mayes machine attached to a Pd system. The Pd system was switched on with the current circulating for 30 minutes before any readings were taken; the fatigue cycle was then begun. Data from the Pd system was exported into an excel spreadsheet and the number of cycles versus crack propagation was then plotted. From these results, values of stress intensity (ΔK) versus crack growth rate were then calculated. In order to calculate the stress intensity factor, a standard equation was used.

$$\Delta K = \frac{\Delta P Y}{B \sqrt{W}} \tag{14}$$

Where, Y = geometry factor, ΔP = change in cyclic load ($P_{max} - P_{min}$), B = sample thickness and W = sample width. Fatigue tests were conducted in fully tension - tension and at a constant frequency.

3.2 Results and discussion

The results of fatigue testing of aluminium honeycomb sandwich panel both in air and in 3.5% sodium chloride solution, are plotted in Figure 22. A total of twenty cyclic deformation tests were conducted in fully tension-tension at a constant frequency of 2HZ, which is equivalent to two cycles per second. The results of fatigue testing of metallic foam sandwich panel both in air and in 3.5% sodium chloride solution, are plotted in Figure 23.

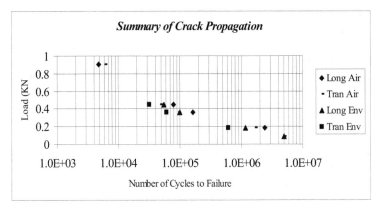

Fig. 22. Aluminium honeycomb fatigue data

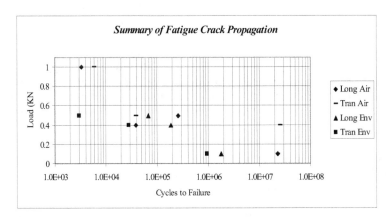

Fig. 23. Metallic foam fatigue data

3.2.1 Aluminium honeycomb

Cyclic deformation data reveals that honeycomb sandwich panel samples do produce consistent results with acceptable repeatability of results even though the honeycomb core is not a conventional structure due to its complex geometry, but because of its homogeneity, it does compare well to the consistent results we would expect from a less complex conventional aluminium solid sample. The results, also, revealed that samples taken from a longitudinal direction constantly have a longer life expectancy, of approximately 40%, then those samples taken from a transverse direction regardless of environmental exposure.

Plotting of fatigue data versus stress intensity for aluminium honeycomb sandwich panels shows that samples tested in a corrosive environment are inferior in performance when compared to samples tested in air. Evidence from crack propagation testing establishes that crack propagation takes place, firstly, within the side plate, leading to some fracture but mainly tearing of the honeycomb structure, only a small amount of crack propagation is evident in the honeycomb structure. The weakest part of the sandwich panel structure appears to be the interface between the aluminium side plate and honeycomb core, with the adhesive used being epoxy resin. Crack propagation testing shows that crack growth is not equal on both sides of the sandwich panel structure; this effect must be due to the complex geometry of the hexagonal core and is a potential difficulty when considering the commercial applications of the aluminium honeycomb sandwich panels.

This research produces a valid method of calculating the Paris exponent, m, with the aluminium honeycomb sandwich panel having a Paris exponent, m, of 1.9. This value is similar when compared to typical values for aluminium alloys of between 2.6 to 3.9.

3.2.2 Metallic foam

Examination of the metallic foam sandwich panel revealed that a consistent form of failure could not be established, with size and position of voids within the metallic foam core having a detrimental effect on failure. Cyclic deformation data revealed that samples tested in air produced inconsistent results showing that the voids within the metallic foam play an important part in crack propagation. However, when samples are tested in an environment, samples taken from the longitudinal direction are superior. This leads to the conclusion that in an environment precipitates within the outer skin have a significant effect on crack propagation. The crack deformation data suggests that due to the complexity of the metallic foam structure and the scatter of results the life of samples exposed to a corrosive environment cannot be correlated with data produced in air. Analysis shows that this is simply untrue and if the two sets of data are plotted a lower and upper trend can be produced, independent of environment and rolling direction, and it is possible to establish a trend of crack growth data within the two bands. This research produced a valid method of calculating the Paris exponent, m. The metallic foam sandwich panel had a Paris exponent, m, of 7.41.

3.3 A proposed method of analysis to predict the fatigue life of sandwich panels

The method of analysis is formed by acquiring the experimental data for aluminium honeycomb and metallic foam sandwich panels. This experimental data is then compared to data produced by calculating the number of cycles to failure. The aim is to calculate the fatigue lives observed experimentally for both aluminium honeycomb and metallic foam sandwich panels. The calculated data will then be used to produce an equation that will predict experimental fatigue life for the complex metallic foam sandwich panels. Conventionally, crack growth rate can be related to the stress intensity factor range using equations (15) and (16).

$$\Delta K = Y\left(\Delta P\right)\sqrt{\pi a} \tag{15}$$

$$\frac{da}{dN} \equiv C(\Delta K)^{m} \tag{16}$$

By rearranging equation (15) and (16) and separating variables and integrating for m ≠ 2 gives,

$$N_f = \frac{2}{(m-2)C(Y\Delta P\sqrt{\pi})^m}\left(\frac{1}{ai^{(m-2)/2}} - \frac{1}{a_f^{(m-2)/2}}\right) \tag{17}$$

Before equation (17) could be solved to calculate residual life from a crack size (a_i) one must know the final or critical crack size (a_f). For the critical crack condition when $a = (a_f)$ equation (15) can be rewritten as;

$$a_f = \frac{1}{\pi}\left(\frac{K_c}{YP_{max}}\right)^2 \tag{18}$$

Where; K_c= Fracture Toughness
The fracture toughness of aluminium honeycomb sandwich panels is 0.91 MPa $\sqrt{m}$ and 0.85 MPa $\sqrt{m}$ for Alulight foam. The calculated crack propagation life versus experimental crack propagation life is shown in Figure 24. In Fig. 24 data can be seen for aluminium honeycomb sandwich panels. The graph clearly shows an excellent correlation between calculated and experimental results.

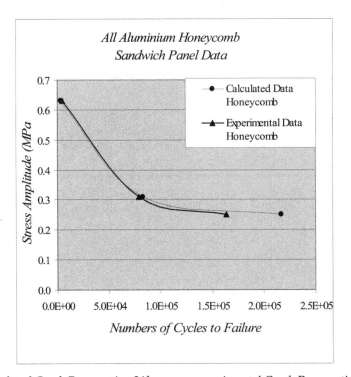

Fig. 24. Calculated Crack Propagation Life versus experimental Crack Propagation Life for Aluminium Honeycomb Sandwich Panel

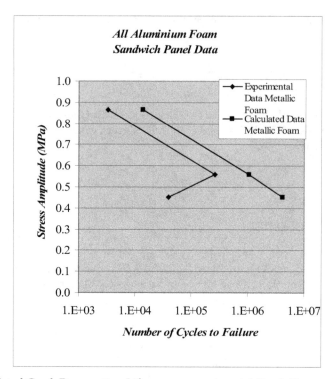

Fig. 25. Calculated Crack Propagation Life versus experimental Crack Propagation Life for Metallic Foam Sandwich Panels

However, the calculated crack propagation life versus experimental crack propagation life for metallic foam sandwich panel is shown in Fig. 25 where it can be clearly seen that calculated data does not correlate with experimental data. The graph illustrates that calculated data always produces a higher number of cycles to failure for metallic foam sandwich panels. The main reason for this is that calculation of stress within the metallic foam structure is complex due to the inhomogenity of the voids within the foam. Equations produced using data from Figs. 24 and 25 were then used to calculate experimental data equation for foam:

Foam Calculated:

$$\sigma = -0.072 \ln N_f + 1.5518 \tag{19}$$

Honeycomb calculated:

$$\sigma = -0.0912 \ln N_f + 1.3589 \tag{20}$$

Therefore, to plot experimental foam data:

$$\sigma = \frac{(-0.1632 \ln N_f - 2.9107)}{2} \tag{21}$$

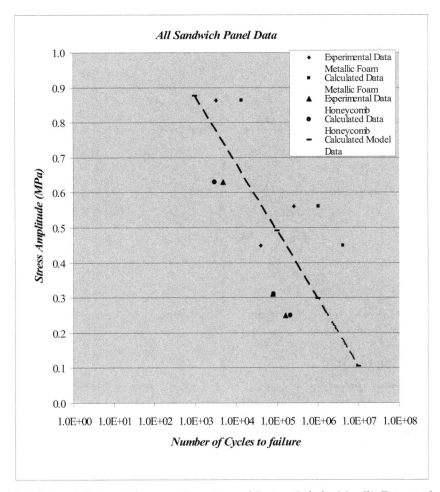

Fig. 26. Calculated Fatigue Life versus Experimental Fatigue Life for Metallic Foam and Aluminium Honeycomb Sandwich Panels

The results of both aluminium honeycomb and metallic foam data for both calculated and experimental cycles to failure is shown in Figure 26. An equation was calculated for each set of data, equations 19 for metallic foam and 20 for aluminium honeycomb respectively. The equations 19 and 20 can then be used to develope an equation 21 to predict the fatigue life of a close cell metallic foam sandwich panel. Using equation 21, calculated life for aluminium honeycomb and metallic foam sandwich panels are compared with original experimental data. What is clear from the Fig. 28 is that all of the experimental data for the two types of specimens correlate with the predicted values. The data correlation proves that a successful model is produced to calculate fatigue life for metallic foam sandwich panels. This model is of extreme importance because it shows that from a structural point of view, metallic foam sandwich panels can successfully replace aluminium honeycomb sandwich panels.

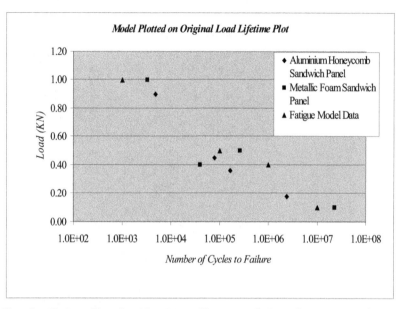

Fig. 27. Showing Fatigue Data for Aluminium Honeycomb, Metallic Foam Sandwich Panel and Calculated Model Data

4. Concluding remarks

The influence of processing conditions in the micromechanical behaviour of Al/SiC composites has been discussed. Two different manufacturing processes (cast and rolled), three reinforcement percentages (20%, 30%, 31%) and three processing states (as received, HT-1, T6 heat treated) have been compared.

The importance of processing conditions in the micro-structural events of segregation and precipitation has been depicted at the micro/nano level using microhardness measurements and nano-scale phase identification of the matrix-reinforcement interface, and the developments of strengthening mechanisms in the composite have been identified. The HT-1 heat treatment condition clearly showed an increase in the microhardness, due to β precipitates as well as other phases and oxides formed in the composite. T6 heat treatment showed the highest microhardness values due to formation of β precipitates, which contribute to strengthening of the interface.

Microhardness and tensile testing results show that the composite micro-mechanical behaviour is influenced by certain factors. In the absence of precipitates (as received state) or in the case of dispersed precipitates (aluminium matrix) the dominant parameters influencing the micromechanical behaviour of the composite are the reinforcement percentage, the interparticle distance and the mean size of particulates. However, when precipitates are concentrated in the areas close to the interface (T6 condition) these precipitates contribute to the strengthening of the composite material.

The thermographic examination of the materials show that heat treated composite samples exhibit regular crack propagation behaviour. Stress concentration, due to the presence of particle reinforcements, produced controlled crack growth and higher stresses, which were related to regular energy release by the material during fracture, indicative of brittle fracture behaviour. On the other hand, the large plastic deformation of the aluminium alloy can be associated with the absence of stress-peaks in conjunction with the monotonic temperature rise for a large part of the temperature / time curve prior to the specimen failure.

A model has been applied to predict the interfacial fracture strength of aluminium in the presence of silicon segregation. This model considers the interfacial energy caused by segregation of impurities at the interface and uses Griffith crack-type arguments to forecast the energy change in terms of the coincidence site stress describing the interface and the formation energies of impurities at the interface. Based on Griffith's approach, the fracture toughness of the interface was expressed in terms of interfacial critical strain energy release rate and elastic modulus. The interface fracture toughness was determined as a function of the macroscopic fracture toughness and mechanical properties of the composite using two different approaches, a toughening mechanism model based on crack deflection and interface cracking and a stress transfer model. The model shows success in making prediction possible of trends in relation to segregation and interfacial fracture strength behaviour in SiC particle-reinforced aluminium matrix composites. The model developed here can be used to predict possible trends in relation to segregation and the interfacial fracture strength behaviour in metal matrix composites. The results obtained conclude that the role of precipitation and segregation on the mechanical properties of Al/SiCp composites is crucial, affecting overall mechanical behaviour.

The tension-tension fatigue properties of Al/SiC composites as a function of heat treatment have been discussed as well as the associated damage development mechanisms. The composites exhibited endurance limits ranging from 70% to 85% of their UTS. The T6 composites performed significantly better in absolute values but their fatigue limit fell to the 70% of their ultimate tensile strength. This behaviour is linked to the microstructure and the good matrix-particulate interfacial properties. In the case of the HT1 condition, the weak interfacial strength led to particle/matrix debonding. In the T1 condition the fatigue behaviour is similar to the HT1 condition although the quasi static tensile tests revealed a less ductile nature.

The crack growth behaviour of particulate-reinforced metal matrix composites was also investigated. Aluminium A359 reinforced with 31% of SiC particles subjected to two different thermal treatments, as well as wrought aluminium 2xxx series specimens, have been examined using thermographic mapping. Heat treated composites, and especially those samples subjected to T6 aged condition, exhibited different behaviour of crack propagation rate and stress intensity factor range than the as-received composite specimens. Furthermore, the composite specimens exhibited different fatigue crack growth rate characteristics than the base aluminium alloy samples. It becomes evident that the path of fatigue crack growth depends on the heat treatment conditions, where crack propagation relies on strengthening mechanisms, such as precipitation hardening. The microstructure of the interphase region was also found to play a significant role in the crack growth behaviour of particulate-reinforced composites. In this sense, T6 heat treated Al/SiCp composite samples exhibits better interphase bonding behaviour than the other composite systems.

The fatigue crack growth curves reveal an approximately linear, or Paris law region, fitting the function $da/dN = C \, \Delta K$. Crack growth rate vs. stress intensity range curves have been obtained using lock-in thermography. These results are in agreement with crack growth rate measurements using the conventional compliance method and calculations based on the

Paris law. It becomes, therefore, evident that lock-in thermography has a great potential for evaluating nondestructively the fracture behaviour of metallic composite materials.

Finally, cyclic deformation data reveals that metallic foam sandwich panel samples do not produce consistent results with acceptable repeatability of results but by using calculated crack propagation life data and experimental data for both aluminium honeycomb and metallic foam sandwich panels a method of analysis has been proposed to predict fatigue life of metallic foam sandwich panels.

5. References

[1] K.G. Kreider, Composite Materials, in: Metallic Matrix Composites, Volume 4, Academic Press, New York and London (1974).

[2] D.J. Lloyd, Particle Reinforced Aluminium and Magnesium Matrix Composites, Int. Mater. Rev. 39, 1-23 (1994).

[3] N.F. Mott and F.R.N. Nabarro, An attempt to estimate the degree of precipitation hardening, with a simple model, Proc Phys Soc 52, 85 (1940).

[4] H.B. Aaron and H.I. Aaronson, Growth of grain boundary precipitates in Al-4% Cu by interfacial diffusion, Acta Meter. 16, 789 (1968).

[5] H.B. Aaron, D. Fainstein and G.R. Kotler, Diffusion-Limited Phase Transformations: A Comparison and Critical Evaluation of the Mathematical Approximations, J. Appl. Phys. 41(11), 4404-4410 (1970).

[6] H.R. Shercliff and M.F. Ashby, A process model for age hardening of aluminium alloys - I. The model, Acta Mater. 38, 1789-1802 (1990).

[7] R.A. Carolan and R.G. Faulkner, Grain boundary precipitation of M23C6 in an austenitic steel, Acta Mater. 36, 257-266 (1988).

[8] S.T. Hasan, J.H. Beynon and R.G. Faulkner, Role of segregation and precipitates on interfacial strengthening mechanisms in SiC reinforced aluminium alloy when subjected to thermomechanical processing, J. Mater. Process. Technol. 153-154, 757-763 (2004).

[9] M. Manoharan and J.J. Lewandowski, In-situ Deformation Studies of an Aluminum Metal-Matrix Composite in a Scanning Electron Microscope, Scr. Metall. 23, 1801-1804 (1989).

[10] M. Manoharan and J.J. Lewandowski, Effect of Reinforcement Size and Matrix Microstructure on the Fracture Properties of an Aluminum Metal-Matrix Composite, Mater. Sci. Eng. A 150, 179-186 (1992).

[11] S.T. Hasan, Effect of heat treatment on interfacial strengthening mechanisms of second phase particulate reinforced aluminium alloy, 14th International Metallurgical and Materials Conference (Metal 2005), Hradec nad Moravici, Czech Republic, (2005).

[12] J.J. Lewandowski, C. Liu and W.H. Hunt Jr., Effects of Microstructure and Particle Clustering on Fracture of an Aluminum Metal Matrix Composite, Mater. Sci. Eng. A 107, 241-255 (1989).

[13] G. Rozak, J.J. Lewandowski, J.F. Wallace and A. Altmisoglu, Effects of Casting Conditions and Deformation Processing on A356 Aluminum and A356-20% SiC Composites, J. Compos. Mater. 26(14), 2076-2106 (1992).

[14] Bitzer T.1997 Honeycomb Technology: materials, design, manufacturing, applications and testing. Chapman & Hall.

[15] Alulight International GmbH

[16] Burman, M and Zenkert D, 1997, Fatigue in Foam Core Sandwich Beams International Journal of Fatigue, Vol 19, Issue 7, pp 551 -561
[17] Shipsha, A, Burman, M and Zenkert, D. 1999, On Mode I Fatigue Crack Growth in Foam Core Materials for Sandwich Panels. Journal of Sandwich Structures and Materials Vol 2, Issuse 2, pp 103-116
[18] Banhart, J and Brinkers, W, 1999, Fatigue Behaviour of Aluminium Foams, Journal of Materials Science Letters, 18:pp 617-619
[19] Olurin, OB, McCullough, Fleck NA and Ashby MF, 2001, Fatigue crack Propagation in Aluminium Alloy Foams International Journal of Fatigue, Vol 23, Issue 5, pp 375-382
[20] Myriounis D. P., S. T. Hasan, N. M. Barkoula, A. Paipetis, T. E. Matikas, "Effects of heat treatment on microstructure and the fracture toughness of SiCp/Al alloy metal matrix composites", Journal of Advanced Materials, vol. 41(3), pp. 18-27 (2009).
[21] Myriounis D. P., S. T. Hasan and T. E. Matikas, "Influence of Processing Conditions on the Micro-Mechanical Properties of Particulate-Reinforced Aluminium Matrix Composites", Advanced Composites Letters, vol. 17(3), pp. 75-85 (2008).
[22] Myriounis D. P., S. T. Hasan and T. E. Matikas, "Microdeformation behaviour of Al-SiC Metal Matrix Composites", Composite Interfaces, vol. 15(5), pp. 485-514 (2008).
[23] Myriounis D. P., S. T. Hasan and T. E. Matikas, "Heat Treatment and Interface Effects on the Mechanical Behaviour of SiC-Particle Reinforced Aluminium Matrix Composites", Journal of ASTM International – JAI, vol. 5(7), published on-line, DOI: 10.1520/JAI101624 (2008).
[24] D. G. Aggelis, E. Z. Kordatos, T. E. Matikas, "Acoustic Emission for Fatigue Damage Characterization in Metal Plates", Mechanics Research Communications", Mechanics Research Communications, vol. 38, pp. 106–110 (2011).
[25] E. Z. Kordatos, D. P. Myriounis, S. T. Hasan, T. E. Matikas, "Monitoring the fracture behavior of SiCp/Al alloy composites using infrared lock-in thermography", Proceedings of SPIE - The International Society for Optical Engineering, Vol. 7294, Article number 72940X, 2009.
[26] Myriounis D. P., E.Z. Kordatos, S. T. Hasan, T. E. Matikas, "Crack-tip stress field and fatigue crack growth monitoring using infrared lock-in thermography in SiCp/Al alloy composites", Strain, published on-line, DOI: 10.1111/j.1475-1305.2009.00665.x (2010).
[27] C. J. McMahon Jr., V. Vitek, Effects of segregated impurities on intergranular Fracture Energy, Acta Metall, 27(4), 507-513 (1979).
[28] L. Shoyxin, S. Lizhi, L. Huan, L. Jiabao, W. Zhongguang, Stress carrying capability and interface fracture toughness in SiC/6061 Al model materials, Journal of materials science letters 16, 863–869 (1997).
[29] X. Q. Xu, D. F. Watt, Basic role of a hard particle in a metal matrix subjected to tensile loading, Acta Metal. Mater., 42(11), 3717-3729 (1994).
[30] H. L. Cox, The elasticity and strength of paper and other fibrous materials, Br. J. Appl. Phys.,3, 72-79 (1952).
[31] J. Llorca, An analysis of the influence of reinforcement fracture on the strength of discontinuously- reinforced metal matrix composites, Acta Metall. Mater., 43(1), 181-192 (1995).
[32] Myriounis D. P., S. T. Hasan, T. E. Matikas, "Predicting interfacial strengthening behaviour of particulate reinforced MMC - A micro-mechanistic approach", Composite Interfaces, vol. 17(4), pp. 347–355 (2010).

[33] Z. Wang, R. J. Zhang, Microscopic characteristics of fatigue crack propagation in aluminium alloy based particulate reinforced metal matrix composites, Acta metal. mater., 42(4), 1433-1445 (1994).

[34] R. G. Faulkner, Impurity diffusion constants and vacancy binding energies in solids, Mater. Sci. Technol., 1(6), 442–447 (1985).

[35] Hasan ST, Beynon JH, Faulkner RG., Role of segregation and precipitates on interfacial strengthening mechanisms in SiC reinforced aluminium alloy when subjected to thermomechanical processing, Journal of Materials Processing Technology 153-154:758-764 (2004).

[36] R. G. Faulkner, L. S. Shvindlerman, Grain Boundary Thermodynamics, Structures and Mechanical Properties, Materials Science Forum, 207-209, 157-160 (1996).

2

Effects of Dry Sliding Wear of Wrought Al-Alloys on Mechanical Mixed Layers (MML)

Mariyam Jameelah Ghazali
Department of Mechanical & Materials Engineering
Universiti Kebangsaan Malaysia
Malaysia

1. Introduction

Aluminium alloys are very attractive compared to other materials like steels, particularly for their mechanical properties. Despite of having a relatively low density (2.7 g/cm³ as compared to ± 7.9 g/cm³ of steel), they also possess high ductility (even at room temperature), high electrical and thermal conductivity and resistance to corrosion and high thermal conductivity. However, aluminium by itself exhibits poor tribological properties and their usage, for example in automotive applications, has been limited by their inferior strength, rigidity and wear resistance, compared with ferrous alloys. With respect to friction and wear behaviour, it has been well understood that the tribological behaviour of aluminium alloys is strongly influenced by the mechanical, physical and chemical properties of the near-surface materials. Whether lubricated or dry sliding, there is evidence that substantial work-hardening occurs at the worn surface. Surface strains can be well in excess of those found in conventional mechanical working. Intimate contact between ductile materials in particular, normally involved transferred materials, which may result in the formation of a mechanically mixed layer (MML). The MML was generally found to be comprised of materials from both contact surfaces, and may also include oxygen, and was known to have very different properties to the Al-alloy. Although the formation of an MML was known to modify wear behaviour, the exact manner was not fully understood. Moreover, very little was known about the effect that matrix alloy composition had on MML formation although it was claimed that the MML could improved wear resistance.

2. Backgrounds

2.1 Sliding wear theory

Wherever surfaces move against each other, wear will occur; damage to one or both surfaces generally involves progressive loss of material (ASM International & 1992 Hutchings, 1992). The rate of removal is generally slow. Although the loss of material is relatively small, it can be enough to cause complete failure of large and complex machinery. Hence, it is essential to develop a thorough understanding of the wear process, especially its mechanism and behaviour, in order to optimise performance. In the current work, only dry or unlubricated sliding wear will be further discussed, even though it is often associated with an environment containing appreciable humidity. When two surfaces slide or roll against each other under an applied load, two forces will exist:

1. The load acts normal to the surface areas that in contact will exert a compressive stress on the materials, which has a similarity with cold working and is usually concentrated in the rolling case.
2. A force exerted by the machine in the direction of motion, overcomes the following resistance:
    *   The friction force, F, that is proportional to the normal load between contacting surfaces.
    *   The static coefficient of friction, that is higher at the start of the motion than the dynamic friction.
    *   Adhesion; the tendency of the two mating metals to adhere to each other. It may result in the surfaces being locally bonded together, forming a junction.
    *   In extreme cases, resistance to motion is caused by abrasive material.

2.2 Wear of aluminium alloys
Due to their low density and excellent corrosion resistance, aluminium has become a substitute for steels especially in structures that require high performance and weight reduction. As with most other metals, aluminium reacts with oxygen in air. A submicron thick oxide layer is formed to provide effective corrosion protection. Aluminium is also non-magnetic and non-toxic, and can be formed by all known metal working processes. The density of aluminium is $2.7 g/cm^3$ or approximately one third the density of steel and aluminium alloys have tensile strengths of between 70 and $700N/mm^2$. At low temperatures the strength of aluminium and its alloys increases without embrittlement in contrast to most steels (Pollack, 1977). Table 1 shows a comparison of the physical characteristics of some of the most important construction materials.

During the 1980's, about 85% of aluminium was used in the wrought form, that is rolled to sheet, strip or plate, drawn to wire or extruded as rods or tubes (Higgins, 1987 and Polmear, 1989). Some of the alloys may undergo subsequent heat-treatment in order to achieve the desired mechanical properties. The most common methods to increase the strength of aluminium alloys are:

*   To disperse any second-phase constituents or elements in solid solution and cold work the alloy; there are known as *non-heat-treatable alloys*.
*   To dissolve the alloying elements in solid solution and re-precipitate them. These are also as heat-treatable or precipitation-hardening alloys (originally known as 'age-hardening' alloys).

	Al	Fe	Cu	Zn
Density (g/cm³)	2.7	7.9	8.9	7.1
Melting Point (YP) (°C)	660	1540	1083	419
Electrical conductivity (%)	63	16	100	30
Specific Heat/Thermal Volume (J/kg, K)	900	450	390	390
Thermal Conductivity (W/m, K)	220	75	390	110
Linear exp. coefficient (10⁻⁶/K)	24	12	16	26
Electrical resistance (10⁻⁹ ohms/m)	27.5	105	17	58
Young's Modulus, (GPa)	70	220	120	93

Table 1. Physical characteristics of some of the most important construction materials

Two major and most common types of wear identified by Eyre (1979) that are relevant to industrial applications of aluminium alloys are abrasive and sliding wear especially for Al-Si alloys. In the case of Al-Si, generally, the hard silicon particles addition will contribute to higher hardness hence increase the wear resistance. Moreover, the particles are surrounded by softer and relatively tough matrix, which then improves the overall toughness of the material. This will lead to wear resistance by favouring more plastic behaviour (ASM Handbook, 1994).

As for aluminum alloys that reinforced with ceramic particles, they have shown significant improvements in mechanical and tribological properties including sliding and abrasive wear resistance (Rittner, 2000). The hard ceramic particles provide protection from further detrimental surface damage. An increase of ceramic hard particles content in alloys may enhance its wear resistance behaviour (Geng et al., 2009). The ageing behaviour of discontinuous reinforced metal matrix composites has been a subject of great interest, which is beneficial to optimise the ageing treatment and providing the experimental and theoretical information for designing the composites properties (Sheu and Lin, 1997). Aluminum nitride (AlN) as a reinforcement material has received much interest in electronic industry because of the need for smaller and more reliable integrated circuit.

For applications, aluminium based alloys have been widely used, for instance Al-Sn alloys as bearing metals in automobile designs. The most important properties of being a bearing metal are that it should be hard and wear-resistant, and have a low coefficient of friction. At the same time, it must be tough, shock-resistant, and sufficiently ductile to allow for *running-in*[1] processes made necessary by slight misalignments.

3. Mechanically Mixed Layers (MML)

3.1 Formation of the MML

In the case of ductile materials like aluminium alloys, most wear mechanism observed are consistent with Archard adhesive wear characterised by plastic ploughing and transfer of material from the counterface. With respect to friction and wear behaviour, numerous authors (Perrin and Rainforth, 1995, Leonard et al., 1997, Jiang and Tan, 1996, How and Baker, 1997 and Rigney, 1998) have concluded that the tribological behaviour is influenced by the mechanical, physical and chemical properties of these near-surface materials. In all cases, a mechanically mixed layer (MML) was present in most dry worn wrought aluminium alloys due to the repetitive sliding. However, significant differences between the MML of each alloy were observed. Their thickness which varied with loads suggested that the subsurface zones of the materials to the sliding and impact wear consisted of 3 zones (Rice et al., 1981) as indicated in Fig. 1.

a. Zones 1 – represents the undisturbed base material or original specimen material in the undeformed state, which experiences elastic deformation and thermal cycling when loaded in tribocontact. Its structure and properties are identical to those prior to the wear test.

b. Zones 2 – consists of the part of the original specimen that has obtained new properties due to repetitive tribocontact. Basically, sufferred deformed intermediate region of the base material. Here, plastic deformation occurs especially in ductile materials, grains are distorted and cracks or voids may nucleate.

---

[1] The process by which machine parts improve in conformity, surface topography and frictional compatibility during the initial stage of use.

c.   Zones 3 – is known as tribolayer which forms in-situ, and usually contains chemical species from the counterface and test environment as well as the bulk material.

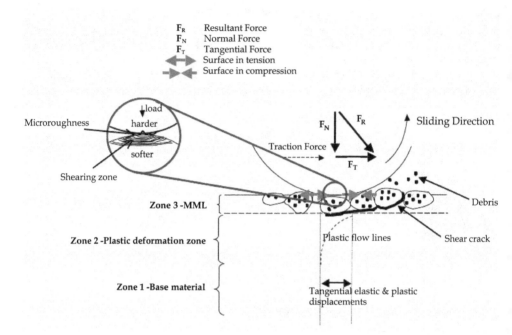

Fig. 1. An illustration of deformation during dry sliding (Ghazali, 2005)

The mixed layer (zone 3), which is commonly known as the mechanically mixed layer (MML), was formed through the incidental mixing between the two materials that statistically occurs at the contact spots under normal pressures. Crack and void formation were generally associated at the zone 2/3 interface and may dictate the dimensions of the wear debris formed (Suh, 1973). The extent and compositional features of these sub-surface zones were found to depend on the conditions of sliding wear, material and environment. Rice et al. (1981-1982) also indicated that these sub-surface zones developed quickly under dry sliding wear conditions. The present work has confirmed that a MML was formed in the sliding wear of the Al-alloys against both counterfaces. Its particles are recognised to have the same physical structure and chemical composition as those of the base pair (Biswas, 2000). The distinctive morphology of the mixed layer has led to a suggestion that its formation was due to a compression of the transfer material and the entrapped debris, which was followed by mechanical mixing during the sliding process. As highlighted by Heilmann et al. (1983), the MML which develops at an early stage (even before loose debris was obtained), is common in both dry and lubricated sliding wear process. In dry sliding condition, a high compressive pressure and large shear strains in the asperities were produced. Heavy plastic deformation and shear strains in the worn surface give rise to dislocation cells and elongated subgrains, as seen in i.e; Fig. 2, which is consistent with Heilmann et al. (1983), Rigney et al. (1981), Chen (1986), Chen and Rigney (1986) and Kuo and Rigney (1992).

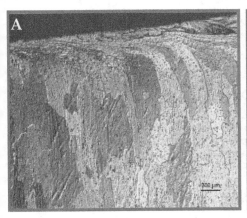 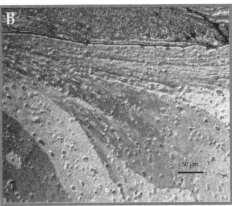

Fig. 2. A is an example of A3004 alloys after 10.8km slid against M2 at 140N. B is the magnified (Normaski) view of a selected area (Ghazali, 2005)

Some MML was very thin and the matrix of the Al alloy almost approached to the top worn surface, is very likely associated with the plastic flow during deformation, where the mixed layer can be replenished with fresh base material by a large plastic flow in the subsurface (Li and Tandon, 1999). Biswas (2000) studied the thickness of the MML appears to be controlled by the depth of crater and abrasion grooves made on the surface. In fact, he further concluded that one of the pre-requisites for the formation of this MML correlates with severe deformation of the top layer of the softer material pair, which in this case is the aluminium alloys. It appears that, when one surface is softer than the other, metal may be transferred from the soft to the hard surface. The material could be transferred back and forth several times during sliding and eventually produce wear debris particles (Heilmann et al., 1983).

3.2 Composition of the MML

In agreement with Rice et al. (1981), Heilmann et al. (1983) and Rigney et al. (1984), the mixed layer is composed of a mixture of two mating materials and from this layer, the loose debris were derived. Based on the results of EDS (Figs. 3), the layer had similar structure and composition to the loose debris. Here, presumably some of the wear debris from the counterface, together with the debris from the pin may have been compacted to form the MML. In other words, it can be concluded that such debris are not derived directly from the base material, with an exception for the case of abrasion, in which microcutting and microploughing are prominent.

In the present study, the MML were found to contain Al, Fe (for Al/M2 case) and O (in the form of oxide), which proves the source of element in the MML obviously originated from the counterface. The oxides were found to be coexisted with other phases in the MML and the wear debris, which is an expected phenomenon since the wear system was exposed to air. They could provide microstructural stability as a second phase in the ultrafine grained structure in debris, as proposed by Rigney et al. (1984). The oxides which have been known to form some protective and some destructive (Fischer, 1997 and Ravikiran et al., 1995) were then fractured and comminuted in further sliding process. The crushed oxides can be dispersed into the mixed surface layer and act as a pinning source of the grain boundaries in the ultrafine mixture in the MML and in the wear debris (Li and Tandon, 1999).

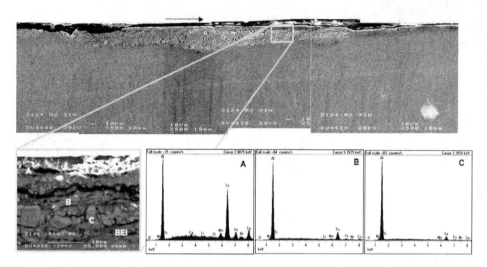

Fig. 3. Subsurface damage of longitudinal cross sections of A2124 alloys against M2 at 140N after sliding 10.8 km. Black arrows indicate the direction of sliding. A corresponding EDS analysis are shown in A, B and C areas (Ghazali, 2005)

3.3 The correlation between the MML and the wear debris

As for wear debris, its formation appeared to occur by two principal mechanisms, namely, the physical displacement of material from the worn surface by the ploughing action of the hard tool steel or alumina asperities, and secondly, delamination of large sheets (up to 1mm in extent) at particularly at high load like 140N. The thickness of the delamination sheets was found broadly consistent with the thickness of the MML, although it could not be defined with certainty whether the delamination occurred within the MML or at the MML/substrate interface. However, the longitudinal cross-sections suggested that both mechanisms were probable. Moreover, the cracks in the MML can give rise to delamination wear as a result of subsurface shear in a manner proposed by Suh (1977) where plate-like wear debris is produced. As one of the main principal wear mechanism in the present study was the delamination of the MML (part or whole), it would be reasonable to expect a correlation between MML thickness and specific wear rate.

3.4 The correlation between MML thickness and specific wear rate

A detailed comparison between several commercial wrought aluminium alloys, namely; A2124, A3004, A5056 and A6092 was carried out for this purpose. For this Al/M2 system, the specific wear rate was relatively insensitive to MML thickness for the A3004 and A5056, although the specific wear rate decreased in a linear manner with increasing MML thickness (refer to Fig. 4a).

In contrast, for the A2124 and particularly the A6092, the specific wear rate was a strong function of the MML thickness. Although a reasonable linear fit was possible for the A2124,

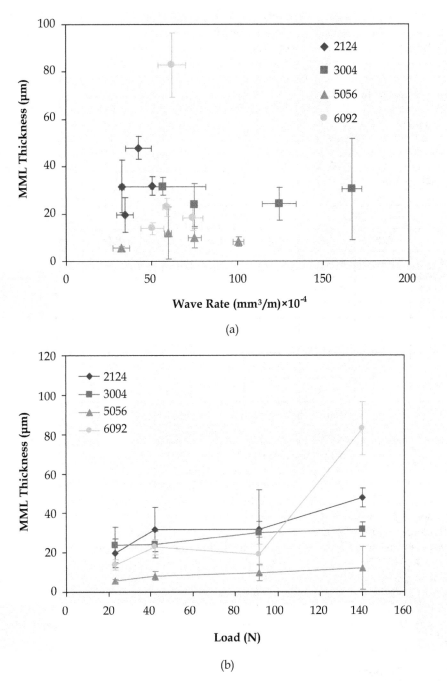

Fig. 4. The relationship between MML thickness and (a) wear rate and (b) load after 10.8km against M2 counterface (Ghazali, 2005)

the A6092 data was better represented by an exponential fit. Since this data did not fit the same trend as the other alloys, the experiment was repeated and measurements re-taken, but with essentially the same result. Thus, the difference in behaviour of this alloy appears to be reproducible. Interestingly, the two alloys where the specific wear rate was relatively insensitive to MML thickness also exhibited MML with the least Fe content (Table 2) and the most homogeneous structure. Conversely, the A6092 exhibited the highest Fe content, the most heterogeneous structure and the greatest influence on the specific wear rate. However, the thickness of the MML cannot explain the dramatic drop in specific wear rate with load observed for the A3004 alloy (Fig. 4b). The thickness of the MML is only one of several potential ways in which the MML can affect wear rate. Clearly for Al/M2 system, the mechanical properties of the MML (in particular hardness and fracture stress) and its adhesion to the substrate are contributing factors.

Element	2124		3004		5056		6092	
	42N	140N	23N	140N	23N	140N	42N	140N
Mg	1.8 ± 0.3	2.0 ± 0.4	-	-	6.8 ± 0.5	6.8 ± 1.1	1.0 ± 0.1	1.7 ± 0.3
Al	85.3 ± 1.9	73.4 ± 5.9	94.9 ± 9.2	95.3 ± 5.6	80.5 ± 7.8	86.8 ± 7.7	61.8 ± 7.6	79.9 ± 5.8
Si	1.0 ± 0.5	-	0.7 ± 0.2	0.8 ± 0.3	1.0 ± 0.4	0.4 ± 0.1	1.6 ± 0.4	1.1 ± 0.3
Mn	1.1 ± 0.1	1.4 ± 0.5	2.5 ± 1.8	1.1 ± 0.2	-	-	0.3 ± 0.1	-
Fe	6.6 ± 1.5	19.1 ± 4.8	1.9 ± 1.5	2.7 ± 1.8	11.7 ± 5.4	6.0 ± 5.5	34.5 ± 7.8	15.9 ± 7.4
Cu	4.3 ± 0.6	4.1 ± 1.1	-	-	-	-	0.8 ± 0.2	1.3 ± 0.6

Table 2. Average quantitative EDS analysis on MML of Al-alloys against M2 (Ghazali, 2005)

In Al/M2 system, the A6092 exhibited the thickest MML and the highest Fe content. Since this was not replicated by the A5056, rather the reverse, it is clear that it is not the Mg content of the A6092 that promotes the formation of a thick MML. Thus, these results imply that stronger adhesion and transfer from the counterface is promoted by the Si in the alloy, while a high Mg content in the Al-alloy reduces adhesion. Similarly, the presence of Cu in the A2124 also appears to have promoted stronger adhesion than an equivalent amount of Mg, although the Cu was not as potent as the Si. The Mn in the A3004 also promoted a relatively thick MML, but one that was more homogeneous than for the A6092. The solubility of these elements in α-Al is in the order Si, Mn, Cu, Mg, which roughly approximates to the thickness of the MML formed. Thus, the observations are in-line with the Archard theory of adhesive wear, as might be expected. However, the level of alloy additions are small (e.g. Si) and it is surprising that the effect was as strong as observed. Thus, the wear performance is largely determined by the properties of the MML.

3.5 Effect of other variables
The atmosphere under an unlubricated wear process can strongly influence sliding wear rates with oxygen content and humidity being probably one of the important factors. In the

case of Al-alloys, it is readily combined with oxygen to form a stable oxide layer. Oxidation, may have opposing effects on the wear process; one, it degraded the surface by removing metal atoms and second, it plays protective role in reducing metallic contact and decrease the wear rate (Degnan, 1995). However, whether or not the environment reaction has a beneficial and detrimental effect on wear rate, it depends strongly on the mechanical interaction of the reaction product with the substrate, particularly under surface plasticity condition, (Rainforth et al., 2002), which is in line with the present work.

Moisture in the environment also can have major effect on wear of metals. Endo and Goto (1978) reported the high humidity had a dentrimental effect on the fretting of aluminium alloys but negligible on carbon steels. Moreover, humidity can control the friction at room temperatures, particularly ceramics as higher coefficient of friction may occurr at temperatures above 800°C.

Beside humidity, all wear processes are influenced by temperature. The temperature reached at the surface of the contact is strongly influenced by the width of the contact (Johnson, 1985) and flash temperature is responsible for many wear and friction effects (Gecim and Winer, 1986). Wear occurs in conjunction with the dissipation of frictional energy in the contact and this is always accompanied by a rise in temperature. The frictional energy is generated by the combination of load and sliding speed and its distribution and dissipation is influenced by other contact conditions such as size and relative velocity. In regards to temperature effects on sample size and mass, contact spots have a tendency to remain in one place much longer on the smaller pin (alloy) than the larger side (counterface), causing stronger local heating in the former. Moreover, in this work, the rotating counterface will mostly experience extra cooling convention than the stationary alloy, which constantly hot due to repeated passage during the test. Local heating at contact spots also has other effect. Most obviously, the local hardness is reduced and thereby the load-bearing area is increased (Kuhlmann-Wilsdorf, 1987).

At high loads like the one used in the present study, 140N, friction heating can induce an increase in temperature, resulting a thermal softening beneath the worn surface, and may affect the wear mechanism (Zhang and Alpas, 1997 and Wang and Rack 1991). Maupin et al., (1992, 1993) studied that large grains were replaced by fine nanocrystalline grains which were relatively free of dislocations underneath the worn surface. Such microstructures could develop only if the temperature of the surface due to friction is very high. In addition, the deformed layer beneath the worn surface could result in higher plastic flow and work hardening resulting in increased wear resistance. At such high temperature, oxidation of the surface is also a possibility, as observed in earlier results.

4. Conclusion

In general, the dry sliding of Al/M2 systems showed the following responses as a result of repeated stress and frictional heat cycle:

- Elements present in the Al-alloy with high solubility in steel promoted a thick mechanically mixed layer, with higher Fe content. The effect was marked even for small contents in the Al-alloy.
- The solubility of these elements in α-Fe is in the order of Si, Mn, Cu, Mg, which roughly approximates the thickness of the MML formed.

- MML with high Fe content tended to be comprised of fragmented particulate, while a low Fe content tended to be associated with a more homogenous MML.
- A linear relationship between specific wear rate and the thickness of the MML was observed for 2124, 5056 and 3004, but not for 6092. The specific wear rate was relatively insensitive to MML thickness for the 3004 and 5056. In contrast, for the 2124 and particularly for the 6092, the specific wear rate was a strong function of the MML thickness.

5. Acknowledgment

The financial support of this research through JPA/SLAB (UKM) program, under the guidance of Professor William Mark Rainforth from University of Sheffield is gratefully acknowledged.

6. References

ASM Handbook Volume 18 (1992), Friction, Lubrication, and Wear Technology (ASM International).

ASM Specialty Handbook (1993): Aluminium and Aluminium Alloys, edited by Davis, J.R., ASM International, ISBN 978-0-87170-496- 2, Materials Park, OH.

Biswas, S.K. (2000). Some mechanisms of tribofilm formation in metal/metal and ceramic/metal sliding interactions. Wear, Vol. 245, No. 1, pp. 178–189.

Chen, L.H. (1986). Deformation, Transfer and Debris for Mation During Sliding Wear of Metals. Scripta Metall., Vol.24, pp.827-83.

Chen L.H., and Rigney, D. A. (1986). Transfer during unlubricated sliding wear of selected metal system. Wear, Vol. 105, pp. 47-61

Degnan, C.C. (1995). The processing and wear behaviour of a W(TiC) reinforced steel matrix composite, PhD Thesis, University of. Nottingham, Nottingham,

Endo K. and Goto H. (1978). Effects of environment on fretting fatigue. Wear, Vol. 48, No. 2, pp. 347-367.

Eyre, T.S. (1979), Treatise on Materials Science and Tech., edited by Scott, D., Academic Press, New York.

Fischer, T.S. (1997). New Directions in Tribology, edited by Hutchings, MEP Ltd, London.

Gecim, B., and Winer, W.O. (1986). Effect of surface film on the surface temperature of a rotating cylinder. ASME J. Tribol., Vol. 108, pp. 92-97.

Geng, L. Zhang, B.P. Li, A.B. Dong, C.C. (2009). Materials Letters 63 pp. 557.

Ghazali, M.J. (2005), Dry Sliding Wear Behaviour of Several Wrought Aluminium Alloys, PhD. Thesis. University of Sheffield.

Heilman, P., Don, J., Suh, T.C., and Rigney, D.A. (1983). Sliding wear and transfer. Wear, Vol. 91, pp. 171-190.

Higgins, R.A. (1987). Materials for the Engineering Technician, 2nd Ed. Edward Arnold, London.

Hutchings, I.M. (1992). Tribology-Friction and Wear of Engineering Materials, 1st edn., Edward Arnold, ISBN 0-340-56184-x, London, UK.

How, H.C., and Baker, T.N. (1997). Dry sliding wear behaviour of Saffil-reinforced AA6061 composites. *Wear*, Vol. 210, pp. 263-272.

Jiang, J.Q., and Tan, R.S. (1996). Dry sliding wear of an alumina short fibre reinforced Al-Si alloy against steel. *Wear*, Vol. 195, No. 1-2, pp. 106-111.

Johnson, K.L. (1985). *Contact Mechanics*, Cambridge University Press, Cambridge.

Kuhlmann-Wilsdorf, D. (1987). Demystifying flash temperatures I. Analytical expressions based on a simple model. *Mat. Sci. Eng.*, Vol. 93, pp. 107-118.

Kuo, S.M., and Rigney, D.A. (1992). Sliding Behavior of Aluminum. *Mat. Sci. Tech.*, Vol. 157, pp. 131-143.

Leonard, A.J., Perrin, C., and Rainforth, W.M. (1997). Microstructural changes induced by dry sliding wear of a A357/SiC metal matrix composite. *Mater. Sci. Tech.*, Vol. 13, No. 1, pp. 41-48.

Li, X.Y., and Tandon, K.N. (1999). Mechanical mixing induced by sliding wear of an Al Si alloy against M2 steel. *Wear*, Vol. 225-229, pp. 640-648

Maupin H.E., Wilson R.D., Hawk J.A. An abrasive wear study of ordered Fe3Al. *Wear*, Vol. 159, pp. 241-247.

Maupin, H. E. Wilson, R. D. and Hawk, J. A. (1993). Wear deformation of ordered Fe-Al intermetallic alloys. *Wear*, Vol. 162-164, pp. 432-440.

Perrin, C. and Rainforth, W.M. (1995). The effect of alumina fibre reinforcement on the wear of an Al-4.3%Cu alloy. Wear, Vol. 181-183, No. 12, pp. 312-324.

Pollack, H.W. (1977). *Mat. Sci. and Met.*, 2nd edn., Virginia, Prentice Hall.

Polmear, I.J. (1989). *Metallurgy of the Light Metals*, 2nd edn., Edward Arnold. New York.

Rainforth, W.M., Leonard, A.J., Perrin, C., Bedolla-Jacuinde, A.,. Wang, Y., Jones, H., Luo, Q. (2002). High-resolution observations of friction-induced oxide and its Interaction with the Worn Surface. *Tribol. Int.*, Vol. 35, pp. 731-748.

Ravikiran, A., Nagarajan, V.S., and Biswas, S.K., Pramila Bai, B.N. (1995). Effect of. Speed and Pressure on Dry Sliding Interactions of Alumina against Steel. *J. Am. Ceram. Soc.*, Vol. 78, No. 2, pp. 356-364.

Rice S.L., Nowothy, H., and Wayne, S.F. (1981-1982). Characteristics of metallic subsurface zones in sliding and impact wear. *Wear*, Vol. 74, pp. 131 -142

Rigney, D.A., ed. 1981, in Fundamentals of Friction and Wear of Materials: Am. Soc. for Metals, Metals Park, Ohio.

Rigney, D.A., Chen, L.H., Naylor, M.G., and Rosenfield, A. (1984). Wear processes in sliding systems. *Wear*, Vol. 100, pp. 195-219.

Rigney, D.A. (1998). Large Strains Associated with Sliding Contact of. Metals. *Mater. Res. Innovat.*, Vol. 1, pp 231-234.

Rittner, M. (2000). *Metal matrix composites in 21st. century: markets and opportunities*, BCC, Inc., Norwalk, C.T.

Sheu, C.Y. Lin, S.J. (1997). *Journal of materials science*, 32 pp. 1741.

Suh, N.P. (1973). The delamination theory of wear. *Wear*, Vol. 25, pp. 111-124.

Suh N.P., Jahanmir, S., Flemming, J.R., Pamies-Teixeira, J.J., Saka, N. (1977). Overview of the Delamination Theory of Wear. *Wear*, pp. 44, 1-16.

Wang A., Rack H.J. (1991). Abrasive wear of silicon carbide particulate—and whisker-reinforced 7091 aluminum matrix composites. *Wear*, Vol. 146, pp. 337-348.

Zhang, J., Alpas, T. A. (1997). Transition Between Mild and Severe Wear in Aluminium Alloys, *Acta Metall.*, Vol. 45, pp. 513–528.

3

Comparison of Energy-Based and Damage-Related Fatigue Life Models for Aluminium Components Under TMF Loading

Eichlseder Wilfried, Winter Gerhard,
Minichmayr Robert and Riedler Martin

Montanuniversität Leoben
Austria

1. Introduction

Thermo-mechanical fatigue is generally due to a cyclic thermal load in conjunction with restrained thermal expansion. Because of the considerable amplitude of strain this load leads to local cyclic plastic deformation and thus to material fatigue. Usually several concurrent and complex damage mechanisms are involved in thermo-mechanical fatigue due to the temperatures and stresses attained. In addition to typical fatigue damage caused by plastic deformation, elevated temperature leads to an increase in corrosive effects (e.g. oxidation) and creep damage. The areas of application are manifold: besides thermo-mechanical fatigue of combustion engine components (e.g. cylinder heads, pistons, exhaust elbows) such effects are common with tanks used in the chemical industry, pipelines, braking systems, turbine blades, as well as all machine tool components and components subjected to elevated operating temperatures. All these applications show cyclic thermal load which, for example, is caused by start-up and shutdown procedures, as well as a mechanical load caused by either restrained thermal expansion (e.g. cylinder heads) or considerable centrifugal forces (e.g. turbine blades).

While in 1992 the maximum specific power for a diesel passenger car was 35 kW/l, the typical ignition pressure was about 130 bar, resulting in a maximum piston temperature of 330 °C. Owing to demands targeting reduction of costs, emissions and fuel consumption, an increase in efficiency by means of "Downsizing" is called for. This is realised by reducing the cubic capacity as well as the number of cylinders and along with additional charging, resulting in an increase in firing pressures and temperature the combustion chamber. The specific power thus obtained is in the region of 70 kW/l, together with ignition pressures of 200 bar and a maximum piston temperature of more than 400 °C, (Reichstein, 2005).

Modern cylinder head materials are typically produced out of aluminium-cast alloys, of which aluminium-silicon-magnesium (AlSiMg) and aluminium-silicon-copper (AlSiCu) alloys are most common. Aluminium and silicon form a eutectic at 577°C and 12 weight percent. The Al-solid solution, silicon and additional secondary phases have a eutectic solidification. The cooling rate influences the dendrite arm spacing (DAS) and the morphology of the eutectic silicon. A high cooling rate leads to a low DAS and finer secondary microstructure. The hypoeutectic alloys are used for cylinder heads and hypereutectic alloys are found in pistons and crankcases.

2. Similarities and differences between LCF and TMF

By their very nature, cyclic thermal loads appear with relatively low numbers of cycles in the low cycle fatigue (LCF) region so that the application of strain-based concepts (e.g. strain life diagrams etc.) is self-evident. If the loading is large enough to produce plastic strain, the number of cycles to failure is relatively low, in the order of less than 10,000 cycles. This total strain predominantly consists of plastic strain, which dominates the fatigue life. Widely used methods to determine the material behaviour are total strain based fatigue tests, whereby the resulting cyclic stress-strain hystereses are investigated. The resulting cyclic stress-strain curves as well as strain S-N curves are the basis for further lifetime evaluation where, depending on the material behaviour, softening or/and hardening effects can be found.

Depending on the application, further influences like temperature, mean strain, strain rate, atmosphere or aging-conditions must be considered. The components are primarily obtained by casting and defects such as pores, shrink holes or oxide inclusions ensued during this process have a negative influence on the lifetime. While these influences are extensively studied for isothermal conditions (Fagschlunger et al., 2006, Oberwinkler et al., 2010, Powazka et al, 2010), scientific understanding of the same for TMF is very limited. While LCF tests are always conducted under isothermal conditions, TMF tests are additionally loaded by thermal cycles, normally defined by a minimum and maximum temperature, dwell time and heating/cooling rates.

As TMF experiments are both very cost-intensive and time-consuming, it is often attempted in practice to estimate the fatigue life of components under thermo-mechanical load by means of more common isothermal LCF experiments. However, this approach may lead to non-conservative fatigue life estimates if the cyclic stress-strain behaviour or the effective damaging mechanisms under TMF loading differ considerably from the material behaviour under isothermal conditions. Furthermore LCF and TMF test results might not correlate due to differing methods used for recording and interpreting the deformation behaviour. In order to avoid misinterpretations it is crucial to pay close attention to the locally and temporally fluctuating temperature field, in particular when recording the TMF deformation behaviour. Thus a fundamental examination of the stress-strain behaviour and the predominant damage mechanisms under TMF conditions is crucial in order to enable accurate fatigue life predictions under thermo-mechanical fatigue loading. This approach can also clarify to which extent the employment of isothermal data is justified (Riedler et al., 2004; Riedler, 2005).

Differences may result from the fact that under TMF loading, as opposed to isothermal LCF loading, during every cycle a broad temperature range is experienced, in which the material properties can change and the material response may differ. The key to a comparison of LCF and TMF data thus lies in the evolution of the microstructure, whose integral behaviour is reflected in the shape of the stress-strain hysteresis loops.

3. Damage mechanisms in thermo-mechanical fatigue

The phenomena during thermo-mechanical fatigue are influenced by a variety of processes within different temperature ranges during a thermal cycle, where especially at elevated temperatures the mentioned damage mechanisms can occur either individually or in mutual interaction. Thus the predominant damage processes are thermally activated gliding of dislocations at low temperatures, cyclic ageing at medium temperatures, and diffusion creep at high temperatures. However, both under IP (temperature and stress cycle are in phase) and OP (temperature and stress cycle are out of phase) TMF loading the microstructure evolution and the oxidation processes are more often than not dominated by the temperature range close to the maximum temperature. The maximum temperature occurs in the tensile stress region in case of an IP-TMF load, and in the compressive stress region in case of an OP-TMF load, see figure 1 (according to Löhe et al., 2004) and figure 2.

On the other hand concerning OP-TMF, crack initiation and growth are linked directly with the processes in the temperature range closest to the minimum temperature where tensile stresses prevail, and are only linked indirectly with processes that occur at the maximum temperature. Nevertheless this indirect influence can be even more distinctive than it would be in isothermal experiments. For example a layer of scale might build up as a result of oxidation which takes effect predominantly at high temperatures. This layer of scale is very brittle at low temperatures and thus causes early crack initiation and accelerated crack propagation under an OP-TMF load.

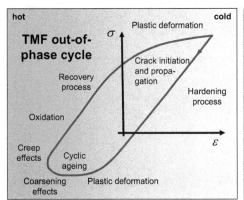
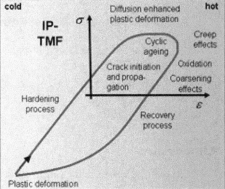

Fig. 1. Active damaging mechanisms during an OP- and IP-TMF cycle (Löhe et al., 2004)

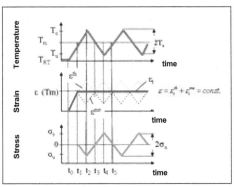
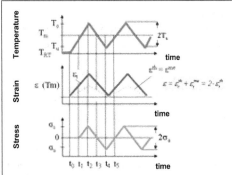

Fig. 2. History of temperature, strain and time under OP-TMF loading (on the left) and IP-TMF loading (on the right)

Crack initiation in various Al-Si-alloys occurs preferentially at the interface between Al-matrix and Si-phase. A meso-scale modelling of the microstructure under thermo-mechanical loading conditions shows stress concentration in the Al-Si-interface (Thalmair, 2009). The repeated recurring thermo-mechanical cycles cause micro-stresses and hence the preferred crack initiation.

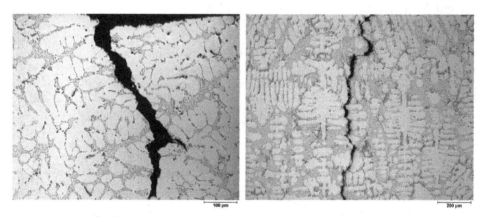

Fig. 3. Crack tip of a TMF-specimen (left) and a thermo-shocked cylinder head (right) of AlSi8Cu3 (Thalmair, 2009)

TMF-test as thermo-shock test of a cylinder head shows very similar crack behaviour, where cracking of the eutectic phase and an inter-dendritic crack-propagation along the interfaces is observed, see figure 3.

4. Methodology

Owing to the multiple effects in components exposed to TMF loading, the developed testing methods are rather varied. Meanwhile, besides LCF experiments and *bi-thermal tests* (Halford et al., 2004) also TMF experiments with combined temperature and strain control are also common. Because of the missing standardisation and the different requirements the experiments differ significantly with respect to heating and cooling temperature application, measurement and control as well as cycle form, strain measurement and consideration of thermal strains. Data from the literature typically allow for little comparability as the information concerning the testing procedures is usually insufficient.

The mechanical loading force application is mostly achieved by means of servo-hydraulic or electro-mechanic testing machines. For this purpose, test equipment manufacturers have started to offer powerful all-in-one systems. However, for special experiments one is still dependent on adaptions or self-made constructions (Riedler & Eichlseder,2004; Minichmayr et al., 2005). Furthermore, it is common to examine specimens with geometries close to the actual component, which only require temperature control. In this case the geometrical constraint is provided for either by the specimen shape or by the mechanical boundary conditions (Simon & Santacreu, 2002; Prillhofer et al., 2005).

It is often tried to simulate the actual behaviour of the component by means of laboratory tests such as for example multiaxial TMF experiments (Otaga & Yamamoto, 2001) or superimposed HCF loading (Minichmayr et al., 2005). By measurement of real components

it is possible to identify actual TMF cycle shapes, which are translated to the test specimen as *industrial cycles* with certain phase shifts between thermal and mechanical strains (Engler-Pinto et al., 1995).
The description of the creep, TMF and LCF testing rigs used for the following experiments as well as a detailed material characterisation can be found in previous papers (Riedler, 2005; Riedler & Eichlseder, 2004; Minichmayr et al., 2005; Minichmayr, 2005).

5. Investigated influences

Firstly, it is important to clarify the governing damage mechanisms that occur in out-of-phase TMF cycles in cylinder heads. Therefore the tests on specimens were specifically designed to take the real circumstances in components as best possible into account – with the aim of using the derived models for lifetime estimations of TMF loaded components. Investigated influences are amongst others (see test matrix in Table 1) mean and local strains, cyclic and constant temperatures, dwell times, pre-aging and aging during service life, HCF-interaction, strain and temperature rates as well as the ratio of mechanical and thermal strain. Further single and multiple step creep tests have been carried out to take into account the stress relaxation phenomena. Additional tests in an argon atmosphere have finally enabled the isolation of the predominating damage mechanism in cylinder heads. All analyses are done in the manner of hysteresis loops, stress-cycle and plastic strain-cycle plots, lifetime diagrams and cyclic deformation behaviour diagrams.

Quasistatic - Creep	LCF	TMF	HCF	Metallographic
Constant temperature	Constant temperature	Maximum temperature	Constant temperature	Fractured surface
Pre-aging	Dwell time	Dwell time	Notch effect	Microstructure
Strain rate	Pre-aging	Pre-aging	Pre-aging	Chem. analysis
Single/multiple step	Mean strain	Mean strain	Dendrite arm spacing	Dendrite arm spacing
Stress relaxation	Strain rate	Temperature rate	Porosity	Porosity
	In lying hole	Rigid clamped - controlled	Mean stress	Precipitations
	Strain amplitude	Strain constraining	Frequency	Grain size
	Argon atmosphere	HCF interaction	Type of loading	Striations
	Incr. step test	Phase shift	Stress amplitude	

Table 1. Test matrix

5.1 Influence of an in lying drilled hole

The aim of this study is to investigate the effects of an in lying drilled hole that is used for an improved quality of the temperature control device, presented in (Riedler & Eichlseder, 2004). The behaviour of the hollow drilled sample is calculated with the finite element method, tested with special LCF test series as well as analyzed by means of fractured surfaces on one wrought and one cast alloy. Whereas the influence on AlCuBiPb is visible, even though, marginally in respect on the lifetime behaviour analyzed with the Manson-Coffin-Basquin (Manson, 1954; Lemaitre & Chaboche, 1985; Basquin, 1910). Approach and the cyclic deformation behaviour analyzed with the Ramberg-Osgood (Ramber & Osgood, 1943) approach, at the Aluminium cast alloy AlSi7MgCu0.5 no difference can be ascertained between the test series of the hollow and solid samples (Riedler & Eichlseder, 2004).

Moreover the analysis by means of fractured surfaces of AlSi7MgCu0.5 of the solid and hollow sample of three LCF strain levels shows assimilable fractured surfaces for each strain level. When decreasing the strain level to lower values, a crack propagation area can be seen beginning at the outside of the specimens. The finite element method shows differences that are of the size of less than one per cent from the maximum axial stress (Minichmayr, 2005).

5.2 Influence of pre-aging

When heat treated aluminium alloys are exposed to elevated or fluctuating higher temperatures in their service life, they show a temperature- and time-dependent aging behaviour which can much decrease the mechanical properties. To investigate these effects on low cycle and thermo-mechanical fatigue, LCF test series at room and higher temperatures, as well as LCF and TMF test series for pre-aged conditions were conducted. Moreover TMF test series with different dwell times at the maximum temperatures were conducted to additionally investigate creep effects.

The first investigation is the separated effect of pre-aging (at an elevated constant temperature) on the deformation and lifetime behaviour by the means of quasi static tests, alternating LCF tests (strain ratio=-1) and temperature-controlled OP-TMF tests (temperature ratio=-1). Figure 4 (left) shows the hysteresis loops for two different total strain levels for non pre-aged and pre-aged specimens at 250°C for 500 hours. At the same LCF strain-level the pre-aged specimens show stress values that are about the half compared to non pre-aged specimens. When investigating the influence of pre-aging on the deformation behaviour by means of tensile tests and LCF tests, at the non pre-aged specimens a high stress hardening tendency can be seen as compared to the tensile test. Pre-aging at 250°C for 500 hours leads to a striking by smaller lifetime in the lower strained LCF region. The deformation behaviour of pre-aged specimens in the manner of stress-cycle or plastic strain-cycle plots shows a nearly straight line without any distinctive hardening or softening, but a markedly higher plastic strain part.

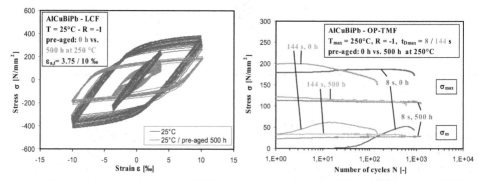

Fig. 4. Influence of pre-aging on the LCF hysteresis loops (on the left) and on the OP-TMF stress-cycle behaviour (on the right)

At the same TMF temperature-level pre-aged specimens at 250°C for 500 hours show the analogous deformation behaviour tendency as obtained at the LCF results, namely a decrease of about 50 per cent compared to non pre-aged specimens, see figure 4 right. The influence of the dwell time decreases with increasing time and temperature of pre-aging.

The differences in lifetime for the dwell time of 8 s and 144 s decreases from a factor more than 6 to a factor of 2, when the specimens are pre-aged for 500 h at 250°C before being tested. After extensive pre-aging the influence of dwell time completely disappears in face of the cyclic deformation and the lifetime behaviour (Riedler & Eichlseder, 2004; Riedler et al., 2005)

5.3 Influence of temperature

A constant elevated temperature influences firstly the quasi static material behaviour and secondly has a time-dependent effect because of hardening vs. softening effects during service life. In this section the time-dependent influence of a constant elevated temperature on LCF is investigated by means of non pre-aged specimens.

A constant elevated temperature of 200°C leads to a higher damage of the material with differences in the lifetime of about one decade compared to the room temperature results. At 250°C the effect is even more drastically, as figure 5 (left) shows. Although at the high strained area the lifetime is a little higher than for 200°C, after the point of intersection at about 100 cycles there is a tremendous drop in the lifetime. Figure 5 (right) shows the summarized presentation of the influences of pre-aging, constant elevated temperature and applied mean strain on the LCF deformation behaviour by means of the plastic strain amplitude part. At a constant temperature of 200°C the stress softening phase starts after a few cycles, what can be seen in an increase of the plastic strain part.

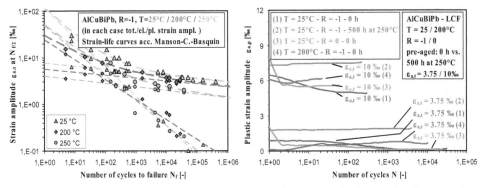

Fig. 5. Influence of elevated temperature on the LCF lifetime (on the left) and influence of pre-aging, elevated temperature and mean strain on the LCF deformation behaviour (on the right)

5.4 Influence of mean strain

Impressed mean strains in the manner of pulsating LCF tests (strain ratio=0) show a visible decrease in lifetime for higher strain levels and only a slight lifetime decreasing effect at lower strain levels. At latest from the half of the number of cycles to failure $N_{f/2}$ on, the cyclic stress deformation curves follow a common progression. Some slight lifetime-decreasing effects ascertainable at pulsating executed test series mostly result from the first few cycles, where the higher tensile stresses and plastic strains (see figure 5 (right)) cause higher damage rates. The comparison of alternating and pulsating executed TMF test series shows the same tendencies as at the LCF results. At latest from $N_{f/2}$ on, the cyclic stress deformation curves follow a common progression and only some slight lifetime-decreasing effects are ascertainable at pulsating executed TMF test series.

5.5 Influence of dwell-time

The typically start-stop-operation of a motor vehicle as well as the alternating fired and non-fired operation causes dwell times. To study this effect out-of-phase TMF tests with four different dwell times at the particular maximum temperatures were conducted (8, 24, 144 and 864 s).

Whereas with the alloy AlCuBiPb, a higher dwell time always causes a lifetime decreasing effect. This is not the case with the alloy AlSi7MgCu0.5, as figure 6 (left) shows, where the strain values are scaled between the minimum and maximum in this range. The TMF strain-life curve for the dwell time of $t_{D3}=144$ s is quite steep for higher strain values (and therefore temperatures). A point of intersection of the curves for the lower dwell times (8 s and 24 s) is visible at about 1000 cycles. This phenomenon is explained with pronounced softening effects in the first few cycles that occur due to the high aging tendency at the high dwell time and temperature level. The capacious over-aging at this level can mainly be seen in the highly plastic parts, what results in an upward movement of the total strain-life curve. For that reason the high over-aging at the dwell time of 144 s shows that a high mechanical strain amplitude can be endured for a longer time compared to the smaller dwell times. If the maximum temperatures are low, this effect turns around at the dimensioning level for cylinder heads (about 5000 cycles) and a lifetime decreasing effect is visible with a higher dwell time.

Two extreme LCF tests were conducted at the mean value for the TMF maximum temperatures. A LCF test series at a constant higher temperature of 250°C and one tested at room temperature, but pre-aged at 250°C for 500 hours. Figure 6 (right) shows that these two LCF strain-life curves span the TMF range for all dwell times of the materials investigated, if the mechanical strain (and not the thermal strain) is considered. The comparison of the LCF hysteresis loops at 250°C with comparable mechanically strained TMF hysteresis loops for all dwell times also shows a good accordance. Moreover the cyclic deformation behaviour according to Ramberg-Osgood shows the best accordance of the LCF-250°C curve with the TMF curves. This investigation shows that the macroscopic behaviour is comparable if the aging status is similar (Riedler et al., 2004).

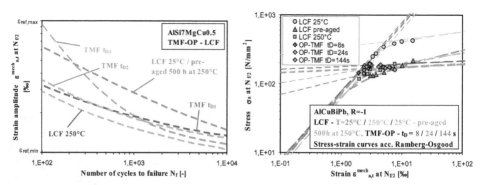

Fig. 6. Influence of TMF dwell-time and LCF-pre-aging and constant elevated temperature on the lifetime (on the left) and on the cyclic deformation behaviour (on the right)

5.6 Influence of phase shifts / strain compensation

In order to investigate the influence of stiffness and phase shifts, different TMF conditions were tested. Besides the ideal OP-TMF situation ($\varepsilon_{t,mech}$=-ε_{th}, K_{TM}=1.0), two overcompensated conditions of the thermal strain (K_{TM}=1.5 and 2.0), a 75% compliance (K_{TM}=0.75, which is near to the real circumstances in cylinder heads) and an ideal in-phase-TMF situation (K_{TM}=-1.0) were tested.

When the local strains are taken into account at rigid clamped specimens, all OP-TMF results can be drawn together in a common strain vs. cycles to failure diagram. Because of creep damage at higher temperatures the IP-TMF lifetime is shorter than the OP-TMF lifetime, as figure 10 shows.

5.7 Influence of strain rates / argon atmosphere

The systematic investigations show that the influence of strain rate on the deformation behaviour is negligible within practical range of temperature rates, but the time and temperature dependent aging behaviour is very important. Unlike the deformation behaviour, the strain rate shows an important influence on the lifetime behaviour due to the additional creep damage involved. These differences and the differences at IP/OP-TMF in the number of cycles to failure allow the separation and investigation of different damage mechanisms. Additional tests in argon atmosphere were executed for this aim, which show a lifetime increasing effect, because of oxidation damage being minimized.

5.8 Influence of HCF interaction

Superimposed HCF-loading has an important influence on the fatigue life of TMF-loaded components. Experiments with superimposed HCF strain amplitudes from 0.01% to 0.1% show a significant decrease in fatigue life depending on the amplitude of the HCF-loads, wherein a small influence of HCF-frequency is found. Metallographic investigations show crack propagation due to HCF-loading and TMF-loading, wherein a combination of HCF-amplitude and the shift in mean-stress causes crack propagation. Good correlation between striations (~crack propagation) and strain amplitude for different loadings is found.

HCF-loadings in a combustion engine occur especially during the heating of the component with maximum ignition pressure. Therefore additional experiments were conducted with superimposed HCF-loadings only during heating period and dwell time. In this case most of the HCF-cycles appear within the compression region of the hysteresis loop. Therefore the effect is obviously reduced. With regard to the typical ignition pressure in a diesel engine, the influence is small compared to the tests without superimposition; see figure 7 (Minichmayr et al., 2005).

5.9 Influence of creep

Because creep effects have to be considered in thermo-mechanical loaded components to take into account stress relaxation phenomena and creep damage, single and multiple step creep tests were carried out. Due to aging effects the decreasing strain rate of the primary creep stage of the single step tests directly merges into the stage of tertiary creep with material softening and therefore increasing creep strain rates. Multiple step tests show, that neither strain nor time are suitable to describe the creep behaviour for that case. Therefore different tests with varying pre-exposure times at test temperature were conducted. The pre-aging time was chosen according to the total time in the multiple step tests. It can be seen, that the minimum creep strain rate of the single step test at 150 MPa is more than 300 times lower compared to the minimum creep strain rate in multiple step test and single step test with pre-aged specimen at the same stress level. Furthermore the test data of the multiple step test and test with pre-aged specimens show a very similar behaviour. Therefore the time at test temperature determines the minimum strain rate independent of strain (Minichmayr et al., 2005).

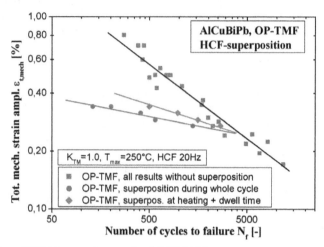

Fig. 7. Influence of HCF interaction on the OP-TMF lifetime

6. Simulation of the cyclic deformation behaviour

6.1 Basics and classification

The description of the elastoplastic deformation behaviour of the construction material forms the basis for assessing the fatigue life of complex components. By means of the finite element method and proper material models it is possible to calculate the local loads (e.g. stress, strain, etc.) under the assumption of adequate boundary conditions.

The most basic material model describes an isotropic plastic hardening independent from the direction of loading. The expansion of the yield surface, which is determined by the *drag stress K* (defined size of the yield surface delimiting the elastic region), can be defined, e.g., in tabular form as a function of the plastic strain. Under cyclic loading each cycle in an isotropic material model leads to further hardening until the maximum strength is obtained and the model shows only elastic, ideal-plastic behaviour. Therefore the isotropic hardening model is adequate only for unidirectional loading. Many materials display the so-called Bauschinger effect under reversed loading (Bauschinger, 1881). The said effect means that plastic deformation occurs already at a significantly lower stress when the load is reversed. The cause of this effect is the formation of dislocation structures, which facilitate plastification in the opposite direction. The Bauschinger effect and the cyclic deformation behaviour, respectively, can only be described by consideration of kinematic hardening, thus using the *back stress α* (which defines the shift of the yield surface in the three-dimensional stress space). In addition high temperatures cause a dependency of stress on the loading rate, which is due to time-dependent processes such as creep. The partitioning of time-independent plastic deformation and time-dependent creep effects for the deformation behaviour at elevated temperatures is already known from Manson (Manson et al., 1971). At the same time this forms the basis of the *strain range partitioning* concept.

A literature review yields a great number of material models which are able to describe the material behaviour for certain kinds of loading. According to (Christ, 1991) they may be classified according to the underlying approach as follows: empirical models, continuum-mechanical models, physically based models and multi-component models.

6.2 Using the ABAQUS Combinend Hardening Model

The cyclic deformation behaviour is described by the ABAQUS® *Combined Hardening Model* dependent on temperature and ageing. The corresponding variables are the temperature-corrected time and the current temperature. The necessary parameters can be correlated with the $R_{p0,2}$ yield limit from the Shercliff-Ashby model (Shercliff & Ashby, 1990). For the calculation of different states of ageing a user subroutine has been developed. On the one hand it allows for an accumulation of the temperature-corrected time on the basis of the temperature-time curve for separate calculation steps, and on the other hand it is possible to calculate directly the state of ageing for a certain ageing time according to the local maximum temperature.

The calculated hystereses and stress-time curves conform very well to the measured experimental data (fig. 8). Under TMF load both the asymmetry of the stress-strain hystereses and the stress relaxation in the dwell time region are expressed correctly.

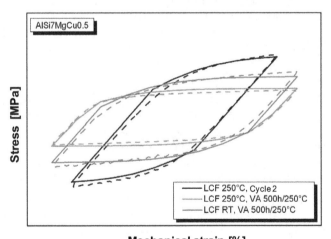

Fig. 8. Comparison of calculated and measured hystereses

7. Simulation of fatigue life behaviour

7.1 Basics and classification

The models available for describing the complex phenomena of thermo-mechanical fatigue range from engineering approaches to physically based models, thereby characterising the combined loading in differing complexity.

Most models for TMF fatigue life calculation are based on linear damage accumulation according to Palmgren-Miner (Palmgren, 1924; Miner, 1945).

Nonlinear cumulative approaches mainly come from Chaboche and Lemaitre (Chaboche & Lesne, 1988; Lemaitre & Chaboche, 1985).

The methods of calculation are mostly based upon local loading parameters such as stress, strain and temperature, which are calculated in complex components by means of elaborate material models. Thus the classification of models describing the fatigue life behaviour under TMF and LCF loading shows the same subdivision as the models describing the cyclic deformation behaviour and shall be briefly explained in the following.

Empirical models

Empirical models represent the major group of this classification. With these models, usually the fatigue life and the parameters of the load cycles are linked. These models are simple, however the partly lacking physical interpretability is a disadvantage.

Furthermore they are effective only for a quite narrow load spectrum as there is no distinction between the individual effects. The empirical models can be set up for various levels of complexity and are divided into approaches based on strain life curves and damage parameters, methods describing creep damage, energy-based approaches, and approaches for partial damage accumulation.

Damage mechanics models

Damage mechanics models usually describe the development of damage by means of methods from continuum mechanics. The origin of these models is found with Kachanov (Kachanov, 1986) and Rabotnov (Rabotnov, 1969) who were concerned with creep damage. Damage mechanics approaches see damage being caused by creep and plastification. As the damage rates are linked to the current damage value, damage accumulation is nonlinear and requires the damage variable to be integrated cycle by cycle.

Physically based models

In general the physically based models play a minor role for practical application, which is due to their complexity and the difficulties in determining their input parameters experimentally. They attempt to characterise the damage development on the basis of atom, vacancy and dislocation movement. At the current state of knowledge and application, physically based models for fatigue life computations are primarily relevant for depicting the physical background of empirical and damage mechanics methods, respectively.

Fracture mechanics models

Fracture mechanics models are linked to the local plastic strains at the crack tip, which can be described, for example, by a ΔJ integral or a modified ΔJ integral, respectively. A link to the physically based models exists with models for micro-crack propagation.

7.2 Application of energetic approaches

An advantage of energy criteria is the significant reduction of parameters in comparison to total strain life curve approaches according to Manson-Coffin-Basquin (Manson, 1954; Coffin, 1954; Basquin, 1910), as energy criteria are able to describe several influences due to the interaction of stress and strain variables. Energy criteria are representative for the cyclic behaviour of materials. They are sound damage indicators which are linked to macroscopic crack initiation and allow for a generalisation to multiaxial loading. The input parameters have to be known, i.e., the local load parameters in consideration of the cyclic deformation behaviour have to be determined in advance. Especially for aluminium alloys the combination of stress and strain variables yields an adequate parameter as the cyclic deformation behaviour is dominantly affected by the ageing effect. If the temperature dependent ageing reduces the stress, the plastic strain increases in a similar way.

Whilst a single-parameter plastic energy approach is an adequate criterion for the highly ductile aluminium alloy AlCuBiPb, for ductile gravity die casting alloys (AlSi7MgCu0.5 and AlSi8Cu3) it is a total strain based energy criterion. For brittle materials such as AlSi6Cu4 lost foam, a fracture mechanics based energy criterion with a cyclic J integral provides the smallest standard deviation (Riedler et al., 2005).

The so-called *unified energy approach* is derived based on this knowledge of material-dependent TMF fatigue life criteria.

$$\Delta W_u = c_u \cdot \Delta W_{u,e} + \Delta W_{u,p} = c_u (\sigma_o \cdot \varepsilon_{a,e}) + (\sigma_a \cdot \varepsilon_{a,p}) \tag{1}$$

$$N_B = A_u \cdot \Delta W_u^{-B_u} \tag{2}$$

The specific hysteresis energy for a representative cycle consists of an elastic and a plastic portion. Whereas the elastic portion is formed by the maximum stress and the elastic strain amplitude, the plastic portion is formed by the amplitude values of stress and plastic strain. The material parameter c_u takes values near 1. The fatigue life is determined by a power law approach according to (2). The quality of computing OP-TMF fatigue life for the examined cylinder head alloy made of aluminium and cast iron by means of the *unified energy approach* can be seen in figure 9. It shows 95% of the data points of the six examined alloys influenced by maximum temperature, average and local strain, pre-ageing as well as ageing in operation lying in a fatigue life scatter band of 2.5, and two-thirds of the data points lying in a scatter band of 1.6.

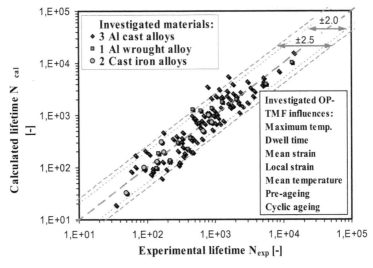

Fig. 9. Quality of the TMF fatigue life simulation by means of the *unified energy approach*

7.3 Application of the damage rate model according to Sehitoglu

The model of Neu-Sehitoglu (Neu & Sehitoglu, 1989) is based on the assumption that overall damage is caused by fatigue, oxidation, and creep:

$$D^{total} = D^{fat} + D^{ox} + D^{creep} \tag{3}$$

or expressed as an equation using the cycles to fracture:

$$\frac{1}{N_B^{total}} = \frac{1}{N_B^{fat}} + \frac{1}{N_B^{ox}} + \frac{1}{N_B^{creep}} \tag{4}$$

The pure fatigue damage portion is described by means of the Manson-Coffin-Basquin approach with the mechanical strain range $\Delta\varepsilon_{mech}$:

$$\frac{\Delta\varepsilon^{mech}}{2} = \frac{\sigma_f'}{E}\left(N_B^{fat}\right)^b + \varepsilon_f'\left(N_B^{fat}\right)^c \tag{5}$$

The parameters E, σ_f', b, ε_f' and c are determined from isothermal fatigue experiments at room temperature. Thereby it is assumed that all experiments at elevated temperature show a similar or shorter fatigue life than at room temperature and furthermore that in these cases the fatigue life reduction is due to the oxidation and creep damage portions.

The oxidation damage portion describes the repeated formation and destruction of an oxide layer at the crack tip as a function of mechanical strain rate, mechanical strain amplitude, temperature and phasing between mechanical strain and temperature:

$$\frac{1}{N_B^{ox}} = \left[\frac{h_{cr}\delta_0}{B\Phi^{ox}K_p^{eff}}\right]^{-\frac{1}{\beta}} \frac{2(\Delta\varepsilon^{mech})^{(2/\beta+1)}}{\dot{\varepsilon}^{1-(\alpha/\beta)}} \tag{6}$$

The temperature dependency of the oxidation is described by means of an Arrhenius approach. The effective oxidation constant is obtained by integration over a complete cycle:

$$K_p^{eff} = \frac{1}{t_C}\int_0^{t_C} D_0 \exp\left(-\frac{Q}{RT(t)}\right)dt \tag{7}$$

The phase factor takes into account that the oxidation damage portion of an OP-TMF load is higher than that of an IP-TMF load:

$$\Phi^{ox} = \frac{1}{t_C}\int_0^{t_C}\phi^{ox}dt \text{ mit } \phi^{ox} = \exp\left[-\frac{1}{2}\left(\frac{(\dot{\varepsilon}^{th}/\dot{\varepsilon}^{mech}+1)}{\xi^{ox}}\right)^2\right] \tag{8}$$

The creep damage portion describes the damage due to pore and intergranular crack formation. The creep damage portion is defined as a function of temperature, equivalent stress, hydrostatic stress and *drag stress*:

$$D^{creep} = \Phi^{creep}\int_0^{t_C} A\exp\left(-\frac{\Delta H}{RT(t)}\right)\cdot\left(\frac{\alpha_1\bar{\sigma}+\alpha_2\sigma_H}{K}\right)^m dt \tag{9}$$

It is made use of the internal variables of the material models according to Slavik-Sehitoglu (Slavik & Sehitoglu, 1987). Creep damage is highest under IP-TMF loading, if the maximum temperature coincides with tensile stress. Isothermal LCF experiments show low creep damage, and for OP-TMF loading it is almost zero. Thus another phase factor is introduced:

$$\Phi^{creep} = \frac{1}{t_C} \int_0^{t_C} \phi^{creep} dt \ \text{mit} \ \phi^{creep} = \exp\left[-\frac{1}{2}\left(\frac{(\dot{\varepsilon}^{th} / \dot{\varepsilon}^{mech} - 1)}{\zeta^{creep}} \right)^2 \right] \tag{10}$$

In (Neu & Sehitoglu, 1989) the parameters are determined by means of elaborate experiments. In doing so, also experiments for measuring the oxide layer growth at different temperatures are conducted. Furthermore the growth of the oxide layer under repeated break-up is also measured.

The parameter adjustment in (Minichmayr, 2005)] is carried out solely by manual parameter variation and automatic parameter optimisation. In addition to the OP-TMF experiments (classical cylinder head applications, basis of energetic approaches) also LCF experiments with different strain rates, LCF experiments in argon atmosphere as well as in-phase TMF experiments were necessary. By means of non-linear parameter optimisation the corresponding model parameters were determined.

The fatigue life computation by means of the Sehitoglu damage model is slightly more accurate in comparison to the energy criteria; 90% of the data points for the cast alloy AlSi7MgCu0.5 lie within a scatter band of 1.85. The biggest advantage results from several damage mechanisms being active at the same time. Likewise it is possible, for example, to predict the in-phase TMF experiments correctly, which are characterised by a dominant effect of creep damage, see fig. 10 and 11.

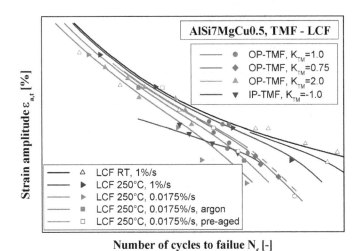

Fig. 10. Fatigue life computing under LCF, OP-TMF, and IP-TMF loading according to the Sehitoglu damage model

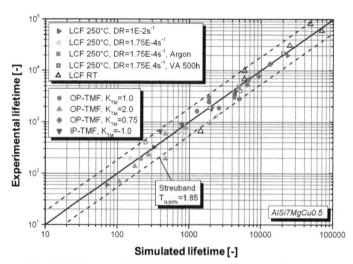

Fig. 11. Quality of the TMF fatigue life calculation using the Sehitoglu damage model

8. Comparative consideration and application to components

Both energy criteria and the model according to Neu-Sehitoglu were used for fatigue life modelling. In contrast to empirical approaches this model distinguishes between different damage mechanisms. The adequate method is chosen according to the type of application; in doing so, it is important to indicate the model limits. Energy criteria seem to represent the best compromise between accuracy and complexity in their application. Major differences occur if the damage mechanisms involved are changing.

Practical application to cylinder heads shows that the fatigue life, calculated on the basis of the dissipated plastic energy, largely depends on the chosen state of ageing. The solution to this problem is an ageing-dependent cumulative damage model.

The commercial fatigue lifetime prediction software FEMFAT (FEMFAT Manual, 2005) features, since version 6.5, a module for calculating the damage under thermo-mechanical loading according to the damage rate model by Neu-Sehitoglu.

With this model it is possible to calculate the local damage portions caused by fatigue, oxidation and creep. The calculation is based on shear strains, which are determined by a critical plane method and are therefore also applicable for multiaxial loading. The predominant part of the overall damage is caused by pure fatigue. The portion of oxidation damage amounts to some 10% in the regions of maximum loading. The fatigue life computation by means of FEMFAT-Sehitoglu provides realistic results concerning the critical areas and fatigue lives.

9. Conclusion

TMF energy criteria is a suitable tool for TMF lifetime assessment of aluminium, provided the limitations of the application are known. They are representative for the cyclic material behaviour and good damage indicators, since they are associated with the macroscopic crack initiation. The damage rate model of Sehitoglu is powerful to describe more influences, albeit with the major disadvantage being the need of an extensive data basis for every specific material. Depending on the application, one specific lifetime calculation method should be preferred.

10. References

Reichstein, S., Hofmann, L. & Kenningley, S. (2005). Entwicklung von Kolbenwerkstoffen für moderne Hochleistungsdieselmotoren, *Giesserei-Praxis*, pp. 380-384, No. 10, Schiele & Schön, Berlin

Fagschlunger, C., Pötter, K., & Eichlseder, W. (2006). Abschätzung der Schwingfestigkeit von porenfreien Randschichten in Al-Gussbauteilen. *MP Materialprüfung*, pp. 142-151, Vol. 48, No. 4, Hanser, München

Oberwinkler, C., Leitner, H., Eichlseder W., Schönfeld, F. & Schmidt, S. (2010). Schädigungstolerante Auslegung von Aluminium-Druckgusskomponenten, *MP Materials Testing*, pp. 513-519, Vol. 52, No. 7-8, Hanser, München

Powazka, D., Leitner, H., Brune, M., Eichlseder, W. & Oppermann, H. (2010). Fertigungsbedingte Einflüsse auf die Schwingfestigkeit von Al-Gussbauteilen, *Giesserei*, pp. 34-42, Vol. 97, No. 7, Gießerei-Verlag, Düsseldorf

Riedler, M.; Eichlseder, W. & Minichmayr, R. (2004). Relationship between LCF and TMF: Similiarities and Varities, *12th International Conference on Experimental Mechanics*, ICEM12, Paper No. 102, Bari

Riedler, M. (2005). TMF von Aluminiumlegierungen – Methodikfindung zur Simulation von thermomechanisch beanspruchten Motorbauteilen aus Aluminiumlegierungen, *Fortschritt-Berichte VDI*, Reihe 5, ISBN 3-18-371805-7

Löhe, D., Beck, T. & Lang, K.-H. (2004). Important aspects of cyclic deformation, damage and lifetime behaviour in thermomechanical fatigue of engineering alloys, pp. 161-175, *Fifth International Conference on Low Cycle Fatigue*, Eds.: Portella, P.D., Sehitoglu, H., Hatanaka, K., DVM, 2004, Berlin

Thalmair, S. (2009). Thermomechanische Ermüdung von Aluminium-Silizium-Gusslegierungen unter ottomotorischen Beanspruchungen, *Dissertation* Univ. Karlsruhe

Halford, G.R., McGaw, M.A.; Bill, R.C. & Fanti, P.D. (1988). Bithermal Fatigue: A Link between Isothermal and Thermomechanical Fatigue, Low Cycle Fatigue pp. 625-637, *ASTM STP 942*, Eds.: Solomon et al., American Society for Testing and Materials, Philadelphia

Riedler, M. & Eichlseder, W. (2004) Temperature control method in elevated and fluctuating temperature fatigue tests, *Materials Engineering*, pp.1-7, Vol. 11, 2004 No. 3, ISSN 1335-0803

Minichmayr, R., Riedler, M. & Eichlseder, W. (2005). Thermomechanische Ermüdung von Aluminiumlegierungen – Versuchstechnik und Methoden der Lebensdaueranalyse, pp. 591-600, *MP Materialprüfung*, Vol. 47, No. 10, Hanser, München

Simon, C. & Santacreu, P.O. (2000). Life Time Prediction of Exhaust Manifolds, pp. 257-267, *Proc. CAMP2002 – High-Temperature Fatigue*, Eds.: Biallas, G., Maier, H.J., Hahn, O., Herrmann, K., Vollertsen, F., Paderborn

Prillhofer, B., Riedler, M. &Eichlseder, W. (2005) Übertragbarkeit von Versuchsergebnissen an Rundproben auf thermomechanisch beanspruchte Bauteile, *1. Leobener Betriebsfestigkeitstage*, Planneralm,

Ogata, T. & Yamamoto, M. (2001). Life Evaluation of IN738LC under Biaxial Thermo-Mechanical Fatigue, pp. 839-847, *Sixth International Conference on Biaxial/Multiaxial Fatigue & Fracture*, Lisboa, Portugal

Engler-Pinto, C.C. Jr., Meyer-Olbersleben, F. & Rézai-Aria, F. (1995). Thermo-Mechanical Fatigue Behaviour of SRR99, Fatigue under Thermal and Mechanical Loading: Mechanisms, pp. 151-157, *Mechanics and Modelling, Kluwer Academic Publishers*, Petten, Eds.: Bressers, J., Rémy, L., Steen, M., Vallés, J.L.

Minichmayr, R. (2005). Modellierung und Simulation des thermomechanischen Ermüdungsverhaltens von Aluminiumbauteilen, Dissertation, Montanuniversität Leoben

Riedler, M. & Eichlseder, W. (2004) Effects of dwell times on thermo-mechanical fatigue, *Zeitschrift Materialprüfung*, Jahrg. 46 11-12, Carl Hanser Verlag, München, S. 577-581.

Bauschinger, J. (1886). Mitt. mech.-techn. Lab., pp. 289, TH München 13, 1886, Zivil-Ing. 27, 1881

Manson, S.S., Halford, G.R. & Hirschberg, M.H. (1971). Creep-Fatigue Analysis by Strain-Range Partitioning, Design for Elevated Temperature Environment, pp. 12-24, Ed.: Zamrik, S.Y., *ASME*, New York

Christ, H.-J. (1991). *Wechselverformung von Metallen*, Springer-Verlag, Berlin

Shercliff, H. R. & Ashby, M. F. (1990). Process Model for Age Hardening of Aluminium Alloys - I. The Model; *Acta metall. Mater*. Vol. 38, No. 10, 1789-1802, 1990.

Palmgren, A. (1924) Die Lebensdauer von Kugellagern, pp. 339-341 *VDI-Zeitung 58*

Miner, M.A. (1945) Cumulative damage in fatigue, Trans. *ASME Journal of Applied Mechanics*, 12-3, pp. A159-A164.

Chaboche, J.L. & Lesne, P.M. (1988). A Non-Linear Continuous Fatigue Damage Model, pp. 1-17 , *Fat. Fract. Engng. Mater. Struct.*, 11

Lemaitre, J. & Chaboche, J.L. (1985). Mecanique des Materieaux Solides, Dunod,

Kachanov, L.M. On Creep Rupture Time, Izv. Akad. Nauk. SSR, Otd Tekh. Nauk, No. 8, pp. 26-31.

Kachanov, L.M. (1986). Introduction to Continuum Damage Mechanics, Martinus Nijhoff Publ., Dodrecht, Holland

Rabotnov, Y.N. (1969). Creep Problems in Structural Members, North Holland Publishing, Amsterdam

Manson, S.S. (1954). Behaviour of materials under conditions of thermal stress, *NACA Report* No. 1170

Coffin, L.F. (1954). A study of the effects of cyclic thermal stresses on a ductile metal, *Trans. ASME 76*, pp. 931-950.

Basquin, O.H. (1910). The exponential Law of Endurance Tests, *Proceedings of the ASTM 10*, pp. 625-630.

Ramberg, W. & Osgood, W.R. (1943). In: *NACA Technical Note* 902.

Riedler, M., Minichmayr, R. & Eichlseder, W. (2005). Methods for the thermo-mechanical fatigue simulation based on energy criterions, pp. 496-503, *6th International Conference of Assessment of reliability of materials and structures: problems and solutions, RELMAS 2005*, St. Petersburg

Neu, R.W.,6 Sehitoglu, H. (1989). Thermo-mechanical Fatigue, Oxidation and Creep, Part II: Life Prediction, *Metal Transactions*, pp. 1769-1783, Vol. 20A

Slavik, D. & Sehitoglu, H. (1987). A Constitutive Model for High Temperature Loading, Part I and II, in Thermal Stress, *Material Deformation and Thermomechanical Fatigue*, Eds.: Sehitoglu, H., Zamrik, S.Y., ASME PVP 123, New York, 1987, pp. 65-82.

FEMFAT: Finite Elemente Methode und Betriebsfestigkeit, *Manual zur Software FEMFAT*, ECS Steyr, 2005.

4

Corrosion Behavior of
Aluminium Metal Matrix Composite

Zaki Ahmad[1], Amir Farzaneh[2] and B. J. Abdul Aleem[1]
[1]Mechanical Engineering Department,
King Fahd University of Petroleum & Minerals, Dhahran,
[2]Department of Metals, International Center for Science,
High Technology and Environmental Sciences, Kerman,
[1]Saudi Arabi
[2]Iran

1. Introduction

Metal matrix composite (MMC) is a material which consists of metal alloys reinforced with continuous, discontinuous fibers, whiskers or particulates, the end properties of which are intermediate between the alloy and reinforcement (Schwartz, 1997). These materials have remained the focus of attention of aerospace, automobile and mineral processing industry because of the several advantages they offer which include high strength to weight ratio, elevated temperature toughness, low density, high stiffness and high strength compared to its monolithic counterpart (the original alloy). The particle reinforced metal matrix composites (PRMMC) satisfy many requirements for performance driven applications in aerospace, automobile and electrical industry. The particle reinforced composites can be tailored and engineered with specific required properties for specific application. The commonly used reinforcing materials are silicon carbide, aluminium oxide and graphite in the form of particles and whiskers. Nominal compositions of some well known alloys which are reinforced with whiskers, fibers or particulate is shown table 1. Figure 1 shows that microhardness increases with an increase in filler content of the composites.

	Si	Fe	Cu	Mn	Mg	Cr	Zn	Ti	Al
Al 6061	0.62	0.23	0.22	0.03	0.84	0.22	0.10	0.1	Bal
Al 7075	0.40	0.50	0.60	0.30	2.5	0.15	5.5	0.2	Bal
Al 6013	0.6	0.50	1.1	0.2	0.8	0.1	0.25	01	Bal

Table 1. Nominal composition of some well known alloys reinforced with whiskers and particles

MMC can be continuous or discontinuous. Discontinuous MMC can be isotropic and can be worked with standard metal working techniques such as extrusion, forging or rolling.

Continuous reinforcement uses monofilament fibers, wires or fibers such can carbon fibers. The reinforcement materials commonly used are graphite SiO_2, SiC, TiC, Al_2O_3 and glasses.

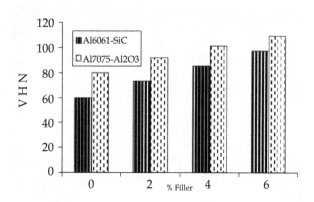

Fig. 1. Microhardness of Al6061-SiC and Al7075-Al2O3 composites (Vaeeresh et al., 2010)

2. Mechanical and physical properties

Metal matrix composites have been shown to exhibit significant improvements in certain physical and mechanical properties over their monolithic metallic counterpart, However, the mechanical properties are strongly dependent on micro structural parameters, in particular, size, shapes volume fraction and orientation of the particles and the composition of matrix.

Parameters	Al 6061	Al 7075	SiC	Al2O3
Flastic Modulus	70 – 80	70 – 80	410	300
Density	2.7	2.81	3.1	3.69
Poisson's Ratio	0.33	0.33	0.14	0.21
Hardness (HB – 500)	30	60	28 W	1175
Tensile Strength(MPA)	115	220	3900	2100

Table 2. Properties of Al 6061 and Al 7075 with and without reinforcement

It is a general observation that the Vickers microhardness observed is greater than the matrix alloy. This is a exemplified by composites, 6061/SiC(p), 6013 SiC(p) and 7075/Al2O3(p) Figure 1 shows the effect of Vol.% of particulates (SiC) on the modulus of elasticity of Al 6061 / SiC, and Al 7075/ Al2O3 composites (Vaeeresh et al., 2010).

The development of metal matrix composites has been a major breakthrough in the last twenty years. The quantum leap in recent years has established their potential for weight critical application in engineering components and structures in aerospace.

It is shown that the tensile strength is increased with increasing volume fraction of SiC particulates. This applies to all metal matrix composites including discontinuously reinforce composite reinforced with SiC particulates or whiskers Figure 2 and 3. The effect of strength may be attributed to the generation of dislocations on cooling of the metal matrix composites. Such dislocations have been observes by TEM. A high dislocation density was observed on Al 6013/SiC (p) interface.

In a TEM experiment, the generation of dislocations started only at 500 K (Vogelsang et al. 1986). It has also been suggested that dislocation were generated in Al – 6061/ 20 SiC MMC below 573 K.

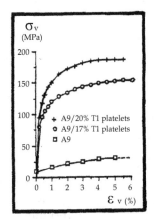

Fig. 2. Effect of the size of the platelets (Massardier et al., 1993)

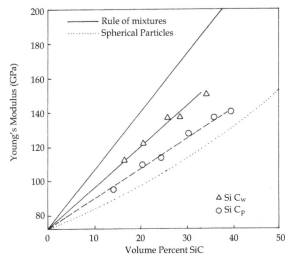

Fig. 3. Young's modulus vs volume percent of SiCw, SiCp and reinforcement (Zaki, 2001)

The elongation (%) of the MMC decreased with increased particulate contents as shown by Al 6061 / 20 SiC (p) – The mechanism of fracture toughness is not fully understood. The presences of large clusters of particles promote crack propagation whereas their uniform distribution retards crack propagation. The fracture toughness values of selected alloys are given in Table 3.

Alloy Designation	Toughness Value
Al2009/ SiC/15(%) W (T 8)	51 mpa √m
Al6061 – 40% SiC(p)	122 mpa √m
Al6013-29SIC(p)	19.5 KSi√2

Table 3. Fracture toughness of selected MMCS

The strains to failure (%) for different Al_2O_3 reinforcement are shown in Table 4 Strain to failure decreases with increase of volume fraction of reinforcement.

Vf %	Percent strain to failure
	(Ef)mm/mn*100
Al 6061	29.26
Al 6061 / 10 vol.% Al2O3	4.72
Al 6061 / 20 vol.% Al2O3	2.29
Al 6061 / 30 vol% Al2O3	1.42

Table 4. Strain to failure of Alloy Al 6061 with increasing volume fractions (Dehlan and Syed, 2006)

Al MMC are finding increasing applications as rotor material in automotive brake systems (Shorowords et al., 2004). Effect of Studies on the effect of sliding velocity on wear friction and tribochemistry of MMC reinforced with 13% SiC or B4C have shown that sliding velocity leads to lower wear rates and lowers friction coefficient for both MMCs.. Studies on interaction between MMC and phenolic brake pads showed that the transfer layer consisting of phenolic pad material acted as a protective layer and reduced wear rates and coefficient of friction. Honda has used aluminum metal matrix cylinder liners in some of their engines including B21Al and H23A, F20 C and F22C.

The effect of cutting speed on tool wear has been investigated. The cutting tool wear increased with increased reinforcement ratios. At constant speed and feed rate, the lowest wear rate has been found in 5 Wt % SiC (p) and the highest wear with 15 Wt % SiC_p increased cutting speed increased the tool wear rate.

From the above description, it may be concluded that the development of MMC has been a big breakthrough in search for stiff high strength materials for aerospace and automotive industry particularly. Whereas the mechanical properties of MMC have remained the focus of attention, the work on corrosion behavior of MMC did not proceed hand in hand with the mechanical and tribological properties. The work on corrosion was undertaken the last decade and a considerable progress has been made in the understanding of corrosion behaviour of metal matrix composites in recent years.

3. Corrosion behavior of Aluminum metal matrix composites

The corrosion behaviour of alloys in sea water 3.5 Wt % NaCl represents an adequate measure of its corrosion resistance. Important results of corrosion studies undertaken in the last decade would be discussed under the following categories.

a. Immersion and long term exposure tests in sea water or 3.5 wt % NaCl.
b. Localized corrosion studies
c. Flow induced corrosion and Erosion corrosion
d. Corrosion inhibition
e. Corrosion mechanism

3.1 Immersion & long term exposure studies

The above studies were conducted in accordance with ASTM designation G 31 – 72 (ASTM, 2004). The results of studies on Al6092 – T6, Al/B4C/20P, Al 6092 – T6 /2oSiC(p), and 6092 – T6 20vol%Al2O3 and monolithic 6061-T6 Al, immersed for 90 days in air exposed 0.5 Na_2SO_4 solution, 3.5 wt% NaCl, ASTM sea water and real sea water were recently described (Hongho et al. 2009). In alloy 6092 – T6 Al/B 4C/20P MMC specimen in ASTM Sea water bubbles were observed. The current over most of the area was found to be anodic. The solution at the anode site was found to be acidic (PH 6.4). Corrosion products were formed as observe after monitoring for three days and the area became more alkaline (PH 8.4). A similar phenomena occurred with alloy 6092 reinforced with 20 Vol. % SiC (p) and gradually the alkalinity increased because of its change of area from and anodic to cathodic. The corrosion rates of MMCS in sea water and ASTM sea water were lower than those in 0.5 M Na_2SO_4 and 3.5 wt % NaCl. The rates of monolithic 6061 – T6 Al in both real and ASTM sea water were significantly lower than those in 3.5 wt % NaCl. The surface morphology after the test showed similar general features, one major feature of the surface morphology was the presence of intermetallic precipitates on the surface. The EDS studies suggested these precipitates to contain Al, Mg, O, and C. Mg and HCO_3 irons as the main species corrosion products.

The formation of precipitates is a greater concern in MMC, as localised corrosion is controlled by the formation of such precipitates. The role of precipitates would be discussed in the relevant section of the paper. In general the corrosion rate of Al MMC decreased with time due to the formation of precipitates.

3.2 Localled corrosion of ALMMC's

If is generally accepted that MMC are in general more prone to corrosion than their monolithic counterparts (Berkely et al., 1998; Turnbull and Corros, 1992; Trzskoma, 1991). Conflicting views have been presented on the causes of the localised corrosion. The results of the studies showed that galvanic corrosion between the matrix and the reinforcement occurs. However, this is related to the machining conditions. Three different machining process; Wielding Electrical Discharge Machine (WEDM), Cemented Carbide Turning and Single Point Diamond Turning were employed for investigation. The test results for different process are shown in Table 5 (Yue et al., 2002).

	E_{Corr} (mV)	E_{pitl} (mV)	E_{Pil} – E_{corr} (mV)	I_{Corr} (Am^{-l})
WED	–761.4	– 633 v	128.4	3.80 TE – 4
Carbide Turning	– 673.6	– 655	186	3.194 E – 2
Diamond Turning	– 928.3	– 655	288.3	1.052 E – 3

Table 5. Electrochemical parameters for different machining conditions (Yue et al., 2002)

The electrical discharge machining showed the highest value of pitting potential. The resolidified layer did not show any extensive pitting. The results show that surface conditions have a major effect on pitting potential and the resistance to pitting may be shown by E_{pit} - $E_{Corrosion}$. The difference above is not sufficient to predict corrosion susceptibility. It may be observed that silicon carbide is an insulator and there is hardly any possibility of cathodic reaction occurring on the surface of particles. The theory that Al/SiC is sensitive to corrosion because of micro galvanic coupling applies to some intermetallic compounds, cathodic to the matrix such as $CuAl_2$ which is formed. So far there is no general agreement on the role of SiC particulates on the mechanism of localized corrosion. The electrochemical behaviour of Al2024/AlSiC has been also investigated by scanning micro reference electrode imaging system (Feng et al., 1981; Isacs & Vyas, 1981). The results of investigations on Al2024/Al SiC (A) are given in Table 6.

Volume Fraction	$E_{pitting}$			$E_{protection}$			$E_{corrosion}$	
	0.01 m NaCl	6.1 m NaCl	0.5 m NaCl	0.01 m NaCl	0.1 m NaCl	0.5 m NaCl	0.5 NaCl	0.1 NaCl
0	– 430	– 497	– 565	– 653	– 620	– 612	– 612	– 574
5	– 460	– 528	– 597	– 750	– 700	– 670	– 670	– 610
10	– 485	– 555	– 625	– 740	– 765115(T)	– 720	– 725	– 688
15	– 538	– 632	– 662	– 700	– 720	– 720	– 750	– 671
20	– 550	– 650	– 692	– 670	– 670	– 775	– 775	– 671

Table 6. Summary of electrochemical data (Feng et al., 1981; Isacs & Vyas, 1981)

It was observe that pitting potential E_p decreased as the volume fraction of SiC particulate reinforcement increased. The relation between the volume fraction and $E_{Protection}$ It was clearly observed that the pitting attack occurred at SiC/Al interface which contained intermetallic Cu and Al precipitates. The presence of Mg, Cu, and Fe compounds in Al6013/20% Vol. of SiC has been confirmed also in another work in recent years (Zaki et al., 2000). The interfacial regions may act as active centers for localized corrosion on immersion in sodium chloride solution. The EDS spectrum of Al_2Cu is shown in Figure 4. The pits on Al 2024/SiC interface are shown in Figure 5. In Al 2024/SiC MMC, Mg may segregate in addition to the precipitates of Al_2Cu Mg and Al_2Cu. The segregated magnesium may form active galvanic couple with Al matrix (Jamaludin et al., 2008). There is also the possibility of the intermetallic precipitates to act as local anodes or cathodes because of the difference between the open circuit potentials of these intermetallic with Al matrix. As seen above the role of the precipitates and inclusion is not clearly understood. However, the evidence of localized corrosion of Al MMC suggests, that the Al/SiC interface in active and responsible for localized corrosion. This is also confirmed by studies on (Al 2009/SiC W) (W = whisker). In the rolled material extensive pitting occurred, and on removing the corrosion products it was observed that the pits contained particles $CuAl_2$ (Rohatgi, 2003). On heat treatment the amount of $CuAl_2$ particles was significantly reduced (Rohatgi, 2003) and the rate of corrosion also diminished which suggested that the heat treatment diminished Mg, Fe and $CuAl_2$ precipitates Figure 6 shows the effect of heat treatment on the corrosion behaviour of T6 and as rolled Al 2009/Sic (w) composite. The corroded surface of as rolled specimens is shown in Figure 6.

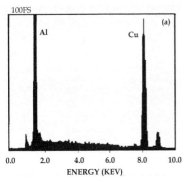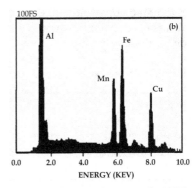

Fig. 4. EDS spectrums of (a) Al2Cu and (b) (CuFeMn) Al6 inclusions (Feng et al., 1998)

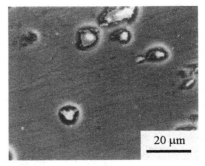

Fig. 5. Scanning electron micrographs of pits on interfaces of (a) SiCp-2024 Al matrix, and (b) inclusions-2024 Al matrix (Feng et al., 1998)

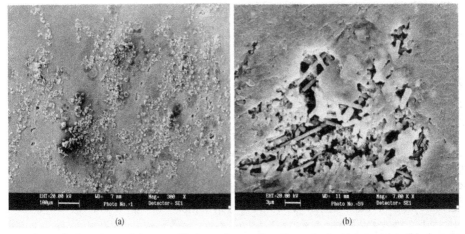

Fig. 6. Corroded surface of the as-rolled specimen after the polarization test (a), (b) showing pit morphology (Yue et al., 2000)

3.3 Flow induced corrosion and Erosion corrosion

The resistance of metallic equipment and structures to the impact of flow induced corrosion is extremely important as it affects their operational life and integrity of equipment. Whereas the effect of velocity on the erosion/corrosion of steel copper, and aluminium alloys are widely reported in literature the information on the metal matrix composite is scanty (Rohatgi, 2004; Griffen &Turnbull, 1994; Lin et al., 1992; Mansfield & Jeanjagnet, 1984; Chen & Mansfeild, 1997; Hihara, 2010; Colman et al., 2011). Studies on Al 6013-20 SiC were conducted in a customized recirculation loop as shown in Figure 7. It consisted of entry valves, a manometer, a centrifugal water pumps, a flow meter and several specimen holders to accommodate flat specimens. Each specimen holders contained four specimens which were housed in an outside container. The velocity was varied by varying the chamber of the specimen holders. Three tempers of Al6013-20 SiC (p) were investigated in the loop. In which a solution of 3.5wt%% NaCl was flowing at velocities ranges from 1-4 ms[-1].

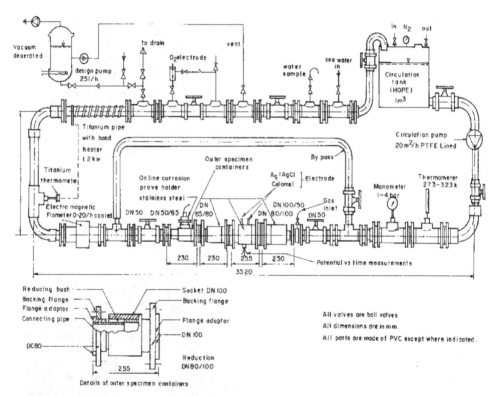

Fig. 7. Schematic diagram of PVC recirculating loop (Zaki, 2001)

After exposure of 100 hours it was shown that temper (0) annealed, and temper F, as fabricated, showed a lower resistance to corrosion in 3.5 wt% NaCl with and without polystyrene suspended particles. Upon increasing the temperature form 30 to 50 and 90 C, the erosion corrosion rate increased as shown in Table 7 and 8 (Zaki, 2007).

Velocity	Corrosion Rate(In 3 weight% NaCl + Vol% Polystyrene(mpy)		
	Temper(0)	Temper (F)	Temper T4
1.0	11.8	9.9	9.6
2.7	12.6	10.8	10.1
3.8	12.9	11.3	11.4

Table 7. Variation of Erosion-Corrosion Rate with Velocity in 3.5wt%% NaCl + 2%Vol Polystyrene (Zaki, 2007)

Temperature (°C)	Velocity	Erosion Corrosion Rate(mpy)		
		Temper(O)	Temper(F)	Temper(T4)
50	1.0	12.1	10.3	9.9
	1.9	3.6	11.2	10.1
	2.7	14.2	12.1	11.4
	3.8	14.9	13.6	10.3
75	1.0	11.9	11.9	117
	1.9	15.5	15.5	161
	2.7	172	17.2	148
	3.8	19.6	19.6	159
90	1.0	13.3	13.3	12.3
	1.9	13.1	15.1	13.6
	2.7	17.8	17.8	163
	3.8	19.7	19.7	17.6

Table 8. The effect of temperature on the erosion – corrosion behavior of Al 6013 – 2051 C (p) in 3.5 wt % NaCl + 2%Vol Polystyrene (Zaki, 2007)

The erosion-corrosion rate increased, linearly with velocity in the presence of SiC particles. It was also found that Temper (T4) of the alloy showed the best resistance to corrosion. The rate of erosion corrosion varied also with temperature. The best resistance offered by T4 may be attributed to the homogenization of the surface structure, less clustering of SiC particles, a uniform distribution of secondary intermetallic phases such as $CuAl_2$ and minimization of micro-crevices (Zaki, 2000). The localized attack was confined to Al 6013/20 SiC (p) interface. A large number of secondary phase particles were observed. After studies showed the presence of Cu 3.55 %, Fe 1.77 %, Mg 1.71 %, and some Cl (0.32%) a high dislocation density was observed at the interface Figure 8 (Zaki, 2000). The formation of coherent films was made more difficult by the protrusion of the particles. This factor adds significantly the erosion –corrosion caused by polystyrene particles. The surface is subjected to a cycle of destruction and reformation of a protective film as a result of impact of polystyrene particles. The corrosion product which accumulates at the interface may act as cathode and increase the cathode / anode area ratio causing an overall increase in the rate of corrosion.

Alloy Al 6013 / 20 SiC (p) in temper T4 offered of temper T4 offered a good resistance to erosion–corrosion. It can be used in water containing Silica or other particulate matter without undertaking any major risk. Al 6013 reinforced with 20 Vol. % SiC (p) was designed to have improved mechanical properties over those of AAl11 6061/SiC (p). The corrosion resistance of al 6013 /20 Sic (p) was determined in fog testing cabinet (Zaki, 2000). A schematic of salt spray chamber is given in Figure 9. The cabinet comprised of a basic chamber level matic test reservoir (1.0 gal salt solution), reservoir (3.0 gal), bubble tank, twin optic fog assembly, and accessories such as a lower assembly bubble tank heater, control valves, and cabinet heaters. The cross section of the assembly is shown in Figure 10. The results obtained for O, F, and T4 Tempers of the alloy composite in the fog cabinet are shown in Table 9.

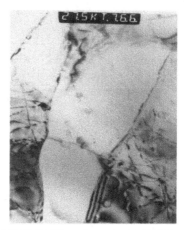

Fig. 8. TEM micrograph of Al 6013/SiC interface showing dislocation generations

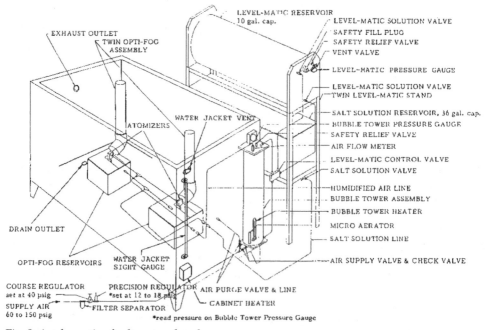

Fig. 9. A schematic of salt spray chamber

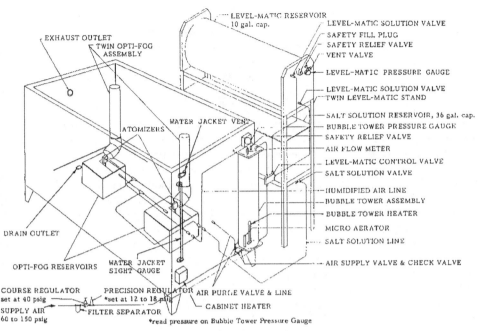

Fig. 10. Cross section of singleton salt fog corrosion test cabinet

Time	Temper-0	Temper-F	Temper-T4
200	10.23 V (19.8)	8.42 (15.78)	7.12 (13.35)
400	9.11 (17.8)	7.78 (14.58)	6.18 (11.09)
600	6.38 (11.96)	6.06 (9.49)	4.38 (8.21)
800	21.92 (9.23)	3.98 (7.46)	2.82 (5.28)
1000	4.66 (8.74)	3.76 (7.05)	2.63 (4.53)
1200	4.27 (8.01)	3.68 (6.90)	2.50 (4.83)

Table 9. Corrosion rates of Al6013/20SiC(p) in Salt Spray Chamber

A decrease in corrosion rate with increased exposure period was observed for all three tempers. The MMC temper T4 showed the highest resistance to pitting. The surface of the composite was often covered with a gelatinous product of aluminum hydroxide $Al(OH)_3$. The pit environment was acidic and bubbles of hydrogen rose from the surface forming corrosion chimneys. The hydrogen bubbles pump up $AlCl_3(OH)_3$ to the outside which reacts with water to form $Al(OH)_3$ (Burleigh et al., 1995). The pitting depth in temper T4 were lower than pitting depths in F and O tempers. It was reported that a high concentration of intermetallic compounds was observed at Al/SiC inter-phases which lead to localized corrosion (A). The corrosion rate of Al6013-20SiC (p) decreased for all tempers on increasing the temperature form 50 to 75°C and increased again on raising the temperature to 100°C. This change may be attributed to the changes brought about by the composition of the protective films from being bayerite (AlO(OH)) to boehmite (Al_2O_3, H_2) as shown by FTIR (Fourier transformation infra-red) spectroscopy).

The corrosion behavior of Al6013–20SiC (p) is a very strong function of Al (OH)3 and once the film formation is completed it becomes independent of oxygen (Beccario et al., 1994). The crystals of boehmite have been observed on the surface of the alloy. The data generated in highly aggressive environment shows promising applications potential of this alloy in salt water and humid environment typical of sea coastal environment in the Gulf Region.

3.4 Effect of Inhibitors

It has been shown in earlier sections that Al/SiC metal matrix composites such as Al 6013-20 SiC (p) exhibit improved mechanical and physical properties compared to wrought alloys. However, they are more susceptible to pitting than their monolithic counterparts (Beccario et al., 1994; Trazaskama, 1990). They also exhibit a higher corrosion rate at velocities greater than the 2.3 ms^{-1} (Zaki, 2000). A variety of surface modification techniques such as anodizing, chromate conversion coatings and organic finishing have been suggested for the protection of aluminum metal matrix composite from localized corrosion (Aylor & Moran, 1986; Lin et al., 1989; Mansfield et al., 1990). Cerium coatings have been the focus of attention in the last decade (Hinton & Arnold, 1986; Davenport et al., 1991).

Studies on to investigate the effect of inhibitors on Al 6013 – 20 SiC (p) included weight loss, Electrochemical and re-circulation loop studies (Zaki, 2009).

Following inhibits solutions were used

a.   1000 ppm $K_2Cr_2O_7$ + 1000 pm $NaHCO_3$ + 3.5 wt % NaCl
b.   1000 ppm Cerium chloride + 3.5 wt% NaCl
c.   1000 ppm sodium molybdate + 3.5 wt % NaCl

The results of inhibitive action of $K_2Cr_2O_7$ + 1000 pm $NaHCO3$ are summarized in Table 10.

Alloy Designation	Velocity (ms-1)	Corrosion rate in mpy(MDD) with no inhibitor	Corrosion rate in mpy(MDD)with inhibitors
Al 6013-20 SiC(p)-O	1.0	11.8(22.1)	3.07(5.76)
	1.9	11.6(21.7)	7.63(14.32)
	2.7	12.9(24.1)	8.4(15.77)
	3.8	13.6(25.5)	9.63(18.08)
Al 6013-20 SiC(p)-F	1.0	9.9(18.5)	3.61(5.68)
	1.9	10.4(19.5)	4.31(8.09)
	2.7	10.8(20.2)	5.53(10.38)
	3.8	11.3(21.2)	6.60(12.39)
Al 6013-20 SiC(p)-T4	1.0	9.6(18.5)	2.01(3.77)
	1.9	10.1(18.2)	2.70(5.07)
	2.7	10.8(20.2)	3.40(6.38)
	3.8	11.4(21.4)	3.80(7.13)

Note; All experiments were conducted in 3.5 wt% NaCl

Table 10. The results of inhibitor action of k2Cr2O7+1000 ppm NaHCO3 (Zaki, 2009)

The reduction in the corrosion rate with $K_2Cr_2O_7$ +NaHCO$_3$ has been attributed to the formation of protective layer of boehmite Al (OH)$_3$, 3H$_2$O and bayrite Al$_2$O$_3$, H$_2$O. The breakdown of the oxide layer leads to pitting. The reduction in the corrosion resistance at increased velocities is caused by continuous removal of protective layer by erodent particles. The protrusion of particulates also makes it difficult to achieve a passivating layer; hence the resistance to the impact of velocity is lowered.

The preferred site for localized corrosion is Al/SiC interface as this site is abundant in intermetallic compound (Zaki, 1998). The existence of thermal stresses and dislocation density at interface affects the kinetics of erosion corrosion and increases the sensitivity if Al/SiC interfaces to erosion-corrosion. Because of the encouraging results of inhibition treatment of Al7057, and Al1000, with cerium chloride and sodium molybdate, studies were further conducted on Al6013 –20 Vol. % SiC(p) MMC. The effect of inhibition treatment is shown in Table 11 below.

Sr. No	Temperature °C	Temper	Corrosion rate in 3.5% NaCl+1000 ppm Mpy(mdd)	Corrosion rate in 3.5% NaCl+1000 ppm Namoo4, CeCl3 Mpy(mdd)
1	50	0	4.72(8.86)	3.8(7.13)
		F	2.24(4.13)	1.8(3.38)
		T4	1.71(3.21)	0.9(1.69)
2	70	0	8.3(15.5)	5.06(9.5)
		F	6.53(12.26)	4.01(7.5)
		T4	2.54(4.77)	2.01(3.72)
3	100	0	12.90(24.2)	8.05(15.11)
		F	11.60(21.7)	8.26(15.41)
		T4	10.19(19.13)	5.41(10.15)

Table 11. Effect of Inhabition Treatment

As shown by table 11 cerium chloride is a more effective inhibitor than sodium molybdate as shown by a larger reduction in corrosion rate brought about by addition of cerium chloride compared to sodium molybdate. The corrosion rate of temper of the MMC is reduced from 19.13 mpy to 3.96 with Cerium Chloride at 100°C which is very significant. Electrochemical studies were also conducted at 50, 70 and 100°C to observe the effect of temperature on inhibition. The electrochemical data obtained by above studies is shown Table 12.

The results of studies summarized in Table 12 clearly established that cerium chloride is a more affective inhibitor than sodium molybdate. The large difference between the corrosion potential (E$_{corr}$) and the pitting potential (E$_p$) shows that the cerium chloride is a more affective inhibitor in 3.5 wt % NaCl. The corrosion potential (E$_{corr}$) shifts closer to E$_p$ which shows the sensitivity of the MMCS to localized pitting in Sodium Chloride without inhibition. The cathodic polarization curve of temper T4 of the alloy in 3.5 wt% NaCl +1000 ppm CeCl$_3$ in dearated condition is shown Figure 11. The curves are overlaid on the main curve. A maximum reduction in current density (from 234 to 25.1uA/cm2) is exhibited by Temper T4 in cerium chloride (Zaki, 2009). The current densities recorded are summarized in the Table 13.

Solution	Temperature °C	Temper	R(K.ohms)	E_corr(mv)	I_corr(μA/cm²)	Corrosion rate mpy(mdd)
Cerium Chloride	50	O	9.004	-0.8	3.727	1.60(2.99)
	50	F	3.301	-0.783	0.57	2.80(5.28)
	50	T.4	2.43	-0.78	1.1	0.47(0.88)
Cerium Chloride	70	O	1.281	-0.909	4.757	2.04(3.81)
	70	F	9.14	-0.915	3.993	1.68(3.14)
	70	T.4	10.07	-0.993	2.155	0.92(1.71)
Sodium Molybdate	50	O	44.5	-0.8	150	1.24(3.22)
	50	F	11.91	-0.716	216.5	1.92(3.60)
	50	T.4	41.3	-0.791	72.8	0.47(0.88)
	70	O	58.77	-0.909	272.6	2.74(5.11)
	70	F	23.76	-0.916	137.0	2.12(4.06)
	70	T.4	34.87	-0.867	111.5	0.75(1.46)
Sodium Molybdate	100	O	100	-0.87	100	8.60(16.00)
	100	F	100	-0.868	100	6.50(12.16)
	100	T.4	58	-0.947	74	3.96(7.41)

Table 12.

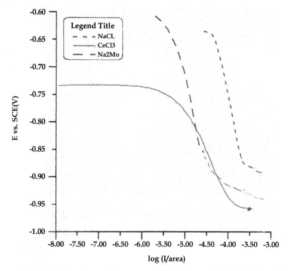

Fig. 11. A cathodic polarization curve of temper T4 of the alloy in 3.5 wt% NaCl + 1000 ppm CeCl₃ in deaerated condition (Zaki, 2009)

Sr. No.	Media	Icorr(μA | cm2)
1	3.5 wt % NaCl	234
2	3.5 wt% NaCl +1000 ppm Cecl3	25.1
3	3.5 wt% NaCl + 1000 ppm NaMoo4	178

Table 13. Current Densities of MMCS after Inhibition (Zaki, 2009)

Cerium chloride acts as a strong cathodic inhibitor for the alloy. Sodium molybdate on the other hand acts as an anodic inhibitor which acts by raising the pitting potential (Up) in the positive direction while maintaining E_{corr} negative to E_p. A typical cyclic polarization curve of the temper T4 of the alloy in 3.5 wt% NaCl + 1000 ppm $NaMoO_4$ is shown in Figure12. The corrosion potential tends to shifts to more positive values.

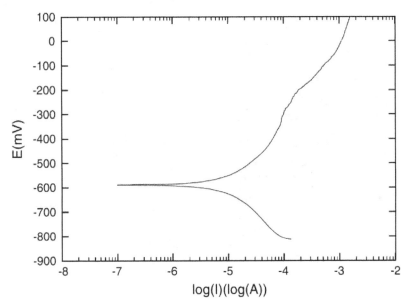

Fig. 12. A typical cyclic polarization curve of Al 6013-20 SiC (p)-T4 temper of the alloy in 3.5 wt.% NaCl + 1000 ppm sodium molybdate in deaerated conditions

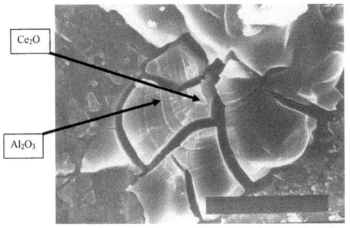

Fig. 13. Surface morphology of Al 6013-SiC (p) in 3.5 wt. % NaCl containing 1000 ppm CeCl3 (Zaki, 2009)

It is interesting to relate the surface morphology to localized corrosion. Typical features of surface morphology after inhibitor treatment are shown in figure 13. Deposition of two types of the particles in concentric rings is seen. These are particles of Ce_2O_3 and Al_2O_3. The square shaped particles of cerium oxide are shown in Figure 14. The oxide layer comprising of Ce_2O_3 and Al_2O_3 are very stable and protect the MMC from corrosion in 3.5 wt% NaCl. However, once the layer reaches a certain thickness, it flakes off. The broken oxide layer in

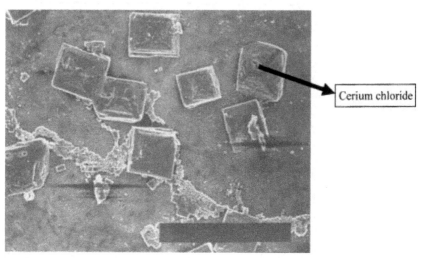

Fig. 14. Square-shaped particles containing predominantly cerium chloride formed on cathodic polarization (Zaki, 2009)

Fig. 15. Broken oxide layer forming blisters (mothballs)

the form of mothballs can be observed in Figure 15. It has been reported that cathodic reaction proceeds at the sites of intermetallic precipitates of copper and its solution becomes alkaline. The film of cerium oxide replaces the film of aluminum hydroxide with increased exposure time (Muhammad & Edwin, 2004; Misra et al., 2007). Whereas the studies on the inhibition of AlMMC are still lacking, there is sufficient evidence to show that cerium chloride is an effective inhibitor for corrosion protection of AlMMC Sodium molybdate is not as effective as cerium chloride shown by the studies reported above composite in chloride containing environment.

3.5 Corrosion mechanism

Despite decades of research no conclusive mechanism on the localized corrosion of Al/SiC(p) composites has been described – The role of intermetallic and dislocation generation at Al/Sic (p) interface has not been conclusively established. No attack a SiC particles has been reported in literature.

From several reliable studies it may be concluded that the pitting potential of monolithic alloys depends on the alloy composition and Ep which is more positive than that of reinforced material (Monticelli et al., 1997; Trazaskoma et al., 1990). The pitting resistance of several MMC investigated followed the order, Al2024 = Al6013 – 20Sic (p) > AL 6061>, Al 6013-20SiC (p) T4=Al5456 (Zaki 2000). In the studies conducted an abundant distribution of copper and secondary phase particles of Mg and Fe were observed.

Copper particles were also present in pit cavities. Analysis of corroded regions at the interface showed a greater concentration of copper compared to the surface away from the interface. The presence of $AlCl_3$ in the oxide film has been indicated by EDS studies (Trazaskoma et al., 1990). Results show a high concentration of copper (3.5%) and Fe (1.77%). There is therefore, a sufficient evidence to show that the increased reactivity at the interface is responsible for localized corrosion of composites. The intermetallic precipitates act as anodic or cathodic sites for initiation of localized corrosion. It is also observed that homogenization of the surface minimizes corrosion the reactivity at the interface is further minimized as shown by temper T4. The SiC particles do not provide any sites for initiation of pits. A higher concentration of copper in pit cavities may be attributed to higher velocities which transport copper ions. Dislocation generation at the interface further activates the interface.

Two more factors are reported to influence, the mechanism of corrosion; Na:YAG laser treatment and machining. Electrochemical studies undertaken showed that the corrosion potential (E_{corr}) increased by 79mv and the corrosion current density decreased by an order of magnitude for the laser treated specimens whereas the untreated surface showed extensive corrosion accompanied by abundant pits. The decrease of corrosion is reported to be due to reduction in the concentration of intermetallic precipitates.

The effect different machining conditions, WEDM, Carbide Turning and Diamond Turning on the electrochemical corrosion behaviour are shown in Figure 16. No significant difference in pitting corrosion potential between the three machining condition was observed. The magnitude of corrosion current for the three machining conditions however differed. Diamond turned specimens showed shallow pits at isolated sites accompanied by a high corrosion rate, whereas Carbide Turned specimens showed extensive pitting because of the hindrance of repassivation of pits due to micro and large crevices present on the surface, pits developed were deeper.

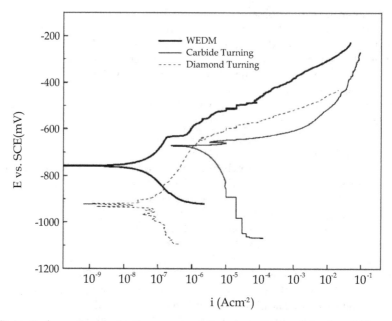

Fig. 16. Potentiodynamic polarization curves of the composite machined to different conditions (Yue et al., 2002)

In the eclectically discharged machined specimen, a resolidified layer of aluminium provided a blanketing effect on the substrate. A reasonable range of passivity was produces on the surface. The high resistance was provided by a layer of oxide on the resolidified aluminium layer created by machining. From the above evidence it is to be understood that surface morphology plays an important role at the Al /SiC interface. Scanning micro reference studies has been employed for direct mapping the active centers on surface of electrode with a low dimensional resolution in micrometers. Evidence of micro pitting has been observed at the open-circuit potentials which were more negative than the pitting potential. In their studies, the same conclusion was reached; i. e the SiC/Al interface is the active center for localized corrosion due to the precipitation of intermetallic compounds (Zaki, 2000).

The observation that heat treatment increases corrosion resistance is shown by temper T4 of Al6013 /20 SiC (p), it is further supported by studies on temper T6 of Al2009/SiC (w), which showed a higher resistance to pitting compared to as rolled specimens It has been already stated above their heat treatment induces homogenization which causes a reduction in the concentration of intermetallic compounds and hence, reduces localized Corrosion.

From the above discussion it may be concluded that the composites are more sensitive to pitting than their monolithic counterparts unless they are subjected to T4 or T6 heat treatment. These is sufficient evidence to show AL/SiC interface is the main target of localized corrosion due to the presence of intermetallic particulates and inclusion which may form micro-galvanic cells and induce localized corrosion. It is also observed that SiC particles are not attacked.

Surface treatment has a significant effect are localized corrosion as shown by the effect of laser treatment and effect of machining on the surface. The scanning micro effloresce electrode studies have shown that Sic/Al interface is the centre f localized corrosion.

Although no conclusion mechanism of localized corrosion of Al MMC exists, there is sufficient evidence to show that Al/SiC interface acts as a centre for localized corrosion and a reduction in the concentration of intermetallic compounds is accompanied by a reduction in localized corrosion as shown by the effect of tempers T4 and T6 on localized corrosion.

4. Conclusion

Based on the studies conducted in the last they decades, the following are the major conclusion on the mechanical and corrosion behaviour of Al MMCs

1. The mechanical properties of fiber, particulate, or whisker reinforced composites are strongly dependent upon the micro-structural parameters, size, shape, orientation and volume fraction of the reinforcement.
2. The tensile strength and Vickers micro-hardness increases significantly with increasing volume fraction of reinforcement as exemplified by Al6013, 6061, 2024 reinforced with particulate and whiskers. The strain to failure also decrease with increased volume fractions of reinforcement - Sliding velocity leads to lower wear rates and lower frication coefficient as shown by SiC and B_4C reinforcements.
3. Increasing cutting speed increased tool wear. The highest wear rate was shown by 15 wt% SiC (p) and the lowest by 5 wt% SiC.
4. Because of accumulation of stress concentration and high dislocation density Al MMC's are sensitive to stress corrosion cracking is 3.5 wt% NaCl – Al6061/20vol% SiC(p)-T6 shows a good resistance to stress corrosion cracking. The polarization curves shifted to higher current densities.
5. The corrosion rates of MMC's decreased with exposure time in long term immersion tests. Heat treatment lowered corrosion rates because of the homogeneous distribution of the precipitates an reduction in their concentration on the Al/SiC interphase electrochemical studies on MMC's showed that the pitting potential decreased with increasing volume fraction of SiC(p) in Al6013 and 6061 reinforce with SiC particulate. Shallow pits contain intermetallic $CuAl_2$.
6. The effect of machining conditions on corrosion showed that electrical discharge machining provided higher resistance to pitting than carbide turning or diamond turning machining.
7. The lowest rate of corrosion was shown by temper T4 if A6013 – 20 SiC(p) is 3.5 wt% NaCl containing silica and other particulate matter.
8. Studies in salt spray chamber showed a good resistance of MMC's to Corrosion and Heat treatment enhances corrosion resistance in corrosive environment.
9. MMC's exhibited a beneficial effect of inhibitor treatment with cerium chloride and sodium molybdate. Cerium Chloride has paved away more effective inhibition than sodium molybdate.
10. The mechanism of corrosion of MMC's is not conclusively understood. It has been, however, shown that the Al/SiC interface us highly reactive due to the presence of intermetallic compounds.

5. References

ASTM; Recommended Practice Designation G 31.72, Standard Practice for Laboratory immersion Testing, (2004) Vol 03, ASTM, Ohio, USA.

Aylor D. M, Moran P. I, Effect of Reinforcement on the pitting Behavior of Aluminum Based Metal Matrix Composites, Jr Electrochem Soc, (1986), Vol 30, p951.

Beccario A.M, Paggi G, Cingaud D, Castellor P, Silicon Carbide Alloy Metal Matrix Composites , Br Jr Corrosion, (1994), Vol 29(1), p65.

Beccario A. M, Poggi G. J, Ginguad D, Castello, Effect of Hydrostatic Pressure on Passivating Powder of Corrosion Layers Formed on 6061-T6 Aluminum Alloy in Sea, Br. Corros Jr, (1994), Vol 29, 1, pp 65-69.

Berkely, D.W, Sallam H.E, Nayeb. Hasemi, H,; The effect of PH on the mechanism of Corrosion and Stress Corrosion and Degradation of Mechanical Properties of AA 6061 and Nextel 440 Fiber-Reinforced AA 6061Composites, Corros Sc,(1998), Vol 40, 2/3 pp, 141-153.

Burleigh T.D, Ludwiczak and Petro R.A, Itergranular Corrosion of an Aluminum Magnesium Silicon Copper Alloy, Corrosion Science, (1995), Vol 5199(1), p50.

Chen C, and Mansfeild F, Corrosion Protection of Al 6092/SiC (p) Metal Matrix Composite, Corr Sc, (1997),Vol 6, PP 1075-108.

Colman S. L, Scott V. D, Enaney M.C, Corrosion Behavior of Aluminum Based Metals Matrix Composites, (2011), Vol 297, 11, DoI: 10.10078/BF001117589, Jr of Mat Sc.

Davenport A. J, Isacs H. S, Kendig M. W; Investigations on the Role of Cerium Compound on the Corrosion Inhibition of Aluminum, Corrs Sc,(1991),Vol 32 (516), p653.

Dehlan Al Khalid; Hafeez Syed, Tensile Failure Mechanism of Al 6061Reinforced with Submicron Al2O3, AJSE, (2006), Vol 31, No 2C.

Feng.Z; Lin, C; LinJ; Lin j; Luo, J; "Pitting Behavior of SiC/2024 Al MMC".

Griffen, A.J; Turnbull, A; 'An Investigation on the Electrochemical Polarization Behavior of Al6061 MMC' Corros Sc (1994), Vol 36,1, 21-35.

Hihara L. H, Corrosion of Metal Matrix Composites, Shriers Corrosion, (2010), Vol 3, pp 2250-2569.

Hinton B. R. W, Dr. Arnold Ryan N. E; Cerium Conversion Coating for Corrosion Protection of Aluminum, Mat. Forum, (1986), Vol 9(3), pp 162.

Hongho Ding, Hawthorn, G.A, Hihara, L.H, Inhibative Effect of Sea Water on the Corrosion of Particulate Reinforce Aluminium Matrix Composites and Monolithic Alloys, Jr. Electrochem Soc, (2009) 156, (100. (35).C159.

Isacs, H.S, Vyas, B; "Electrochemical corrosion Testing, ASTM, STP 727, 1981, p, 3.

Jamaludin, S.B, Yusoff Z, Ahmed R.R, Comparative Study of Corrosion Behavior of A.A. 2009/20 Vol% SiC(w), Porpugaliay Electrochimica Acta, (2008), Vol 26, pp 291-301.

Lin, S; Shih, H; Mansfield F; Corrosion Protection of Aluminium Alloys and Metal Matrix Composites by Polymer Coatings; Corros Sc (1992), Vol 23, 9, pp 1331-1349.

Lin S, Greene H. Shih and Mansfeld F, Corrosion Protection of Al Metal Matrix Composites, Corrosion, (1989), Vol 45(8), p615.

Mansfield S, Lin S, Kim H, Shih, Pitting and Passivation of Al Based Metal Matrix Composites, J Electrochem Soc, (1990), Vol. 137, pp 75-82.

Mansfield F. S.L. Jeanjagnet; The Evaluation of Corrosion Protection Measures For Metal Matrix Composites, Corros Sc (1984), Vol 26, pp 727-734.

Misra, Ajit Kumar, Balsubramanium, Corrosion Inhibition, Material Chemistry and Physics, (2007), Vol 103, 2, 3, pp 385-393.

Monticelli C, Zucchi C, Bruuonoro, Trabanelli C, Stress Corrosion Cracking Behaviour of some Aluminum Base Metal Matrix Composites, Corrs Sc, (1997), Vol 39,10, pp. 1949-1063.

Muhammad Ashraf, P. Leela Edwin, Evaluation of Corrosion Inhibition by Cerium on Aluminum under Marine and Laboratory Environment, 2nd Jr of Chem. Tech, (2004), Vol 11, pp 672-677.

Rohatgi, P.K; 'Aqueous corrosion of Metal Matrix Composites, Comprehensive Composite Materials, (2003), Chapter 3.18, , pp 481-500, Elsevier

Schwartz, M.M "Composite Material Processing Fabrication and Applications" Prentice Hall, USA (1997).

Shorowords,K.M; Haseeb, A.S.M.A; Celic,j.P; "Studies on the wear Friction and Tribochemistry of MMC Sliding against Phenolic Brake Pads, Wear" (2004), pp1176-1181

Turnbull, A.Br, Corros Jr 1992, Vol 27, p.27-35.

Trzskoma, P.P; in "Metal Matrix Composite Mechanism and Properties" Academic Press, (1991), p, 383.

Trazaskama P. P, Pit Morphology of Aluminum Alloy in Silicon Carbide Alloy Metal Matrix Composites, Jr Corrosion, (1990), Vol 46, p402.

Trazaskoma P.P, Maccefferty E, Crowe C.R, Corrosion Resistance of Al Based Metal Matrix Composites, Corrs Sc, (1990), 46, p402.

Vaeeresh Kumar, G.B., Rao, C.P; Selvararj.N; Bhagya Shekar, M.S.B; Studies on Al 6061-SiC and Al 7075-Al2O3 Metal Matrix Composite; Jr for Mater and Mater characterization and Eng, Nov (2010), Vol 9, pp, 43-55.

Vogelsang, M; Arsenault, R.J; Fisher, R.M; In SituHVEM study of Dislocation Generation of Al/SiC Interface in Metal Matrix Composites, Met.Trans A, (1986), Vol 17A, p139.

Yue, T.M; Yu, J.K; Maki, H.G; "Corrosion Behavior of Aluminium 2009/SiC Composite Machined to Different Conditions". Jr Mat Sc Letters, (2002), Vol 21, 14, pp1069-1072

Yue, T.M; Wu, Y.X; Man, H.c; 'On the Role of CuAl2 Precipitates in Pitting Corrosion of Aluminium (2009), SiC Metal Matrix Composites' Jr of Materials Sc (2000), Letters 9, pp 1003-1006.

Zaki Ahmad and Abdul Aleem B. J, The Effect of Inhibitors on the Susceptibility of AL 6013, SiC Interface to Localized Corrosion, Jr of Mat Eng, Perf,(2009),Vol 18,2, pp129-136.

Zaki Ahmad, Mechanical Beauvoir and the Fabrication Characteristics of Aluminum Metal Matrix Composites, Jr of Reinforced Composite Material, (2007),Vol 1, 4, pp 3027-3033.

Zaki Ahmad, Abdul Aleem. B.J; Degradation of Aluminium Metal Matrix Composites in Salt Water and its Control, Mater and Design, (2001), Vol 23, pp173-180.

Zaki Ahmad, Paulette, P.T; and Aleem B.J.A; "Mechanism of Localised Corrosion of Aluminium Silicon Carbide Composites in a Chloride Containing Environment' Jr. Mat Sc (2000), 35, pp 2573-2579.

Zaki Ahmad & Abdul Aleem B. J, Corrosion Resistance of a New Al 6013-20 SiC in Salt Spray Chamber, Jr of Mat SC and Eng, (2000), Vol 9, 3l, p338.

Zaki Ahmed & Abdul Aleem B.J, Corrosion Resistance of New Aluminium Al 6013/20SiC(p) in Salt Spray Chamber,Jr Mat Sc and Eng,(2000),Vol 9(3),p338.

Zaki Ahmed, Paulette, P, T, Aleem B.J.A, Mechanism of Localize Corrosion of Aluminum Silicon Carbide Composites in Chloride Containing Environment, Jr Mater Sc,(2000),Vol 3, 5, pp2573-2579.

Zaki Ahmad & Abdul Aleem B. J, The Erosion Corrosion of Al -SiC Composites in Sea Water, (1998), Final Report, KACST, 14-65, King Abdul Aziz City of Science and Technology, Riyadh, Saudi Arabia.

Lima, P.; Bonarini, A. & Mataric, M. (2004). *Application of Machine Learning*, InTech, ISBN 978-953-7619-34-3, Vienna, Austria

Li, B.; Xu, Y. & Choi, J. (1996). Applying Machine Learning Techniques, *Proceedings of ASME 2010 4th International Conference on Energy Sustainability*, pp. 14-17, ISBN 842-6508-23-3, Phoenix, Arizona, USA, May 17-22, 2010

Siegwart, R. (2001). Indirect Manipulation of a Sphere on a Flat Disk Using Force Information. *International Journal of Advanced Robotic Systems*, Vol.6, No.4, (December 2009), pp. 12-16, ISSN 1729-8806

5

Interrelation Between Failure and Damage Accumulation in the Pre-Fracture Zone Under Low-Cycle Loading

Vladimir Kornev, Evgeniy Karpov and Alexander Demeshkin
Lavrentyev Institute of Hydrodynamics SB RAS
Russia

1. Introduction

When structures are loaded in catastrophic mode of operation, localization of irreversible strains occurs in regions of stress concentration. This is caused by the geometry of a structure or by the presence of fetaures (dints, holes, cracks, inclusions with mechanical characteristics different from properties of original material). If loading is in progressing, repeated loads cause gradual degradation of the material in regions of the localization of inelastic strain. As a result, this leads to generation and extension of cracks and to loss of a load capacity of the structure. Study of regularities of material degradation in regions of strain localization will permit one to appreciate possible structure resources in a catastrophic situation or consequence of failure, which affects subsequent behaviour of a structure in a common regime.

The distinctive features of fracture surface microrelief of metallic components in fatigue are fatigue striations oriented normally to the crack extension direction. It is appropriate to relate formation of fatigue striations to stepwise crack tip advance, and to record residual deflection of a beam under three-point bending when the loading corresponds to the low-cycle fatigue. A current striation may be formed after several loading cycles due to arrest of a fatigue crack after each advance of its tip in the Laird- Smith model (Laird & Smith, 1962), the material being embrittled in a pre-fracture zone at each loading type. Damage accumulation in the pre-fracture zone is associated with accumulation of inelastic strains in this zone.

The aim of the present work was to study damage accumulation in the regions of inelastic strains near the notch tip having a finite width. Two cases are considered: i) symmetric three-point bending of a beam (the edge notch is made on the underside of a specimen in the transverse symmetry plane, damage accumulation is estimeted by the increment of a residual deflection); ii) tension of a plain specimen with a narrow edge notch (direct viewing of fatigue crack propogation was performed using digitized microscope with resolution of about 22500 pxel/mm^2). Mechanisms of deformation, damage accumulation and failure of material under fatigue conditions have been proposed.

In the first case, the choice of geometry of a specimen and the loading type are governed by the following considerations.

Under symmetric three-point bending of a beam with the transverse notch, main inelastic strains are concentrated ahead of the notch tip where stress concentration occurs. Change of mechanical characteristics of material in this region under repeated loading conditions causes the increase in residual deflection of the specimen. Thus, the amplitude of the residual deflection can be used as a measure of damage accumulation in the zone of localization of inelastic strains. This provides, from macroscopic phenomena, a possibility for qualitative and quantitative estimating the changes directly exhibited by the material due to processes of fatigue fracture. Moreover, in this case, the notch can be considered as a model of an edge crack with the blunted tip, and the region of localization of inelastic strains can be considered as a pre-fracture zone ahead of the tip of this crack.

In the second case, consider a plain specimen with one edge notch since two symmetrical notches lead to uncertainty in a choice of the point of crack initiation. Besides, after a crack initiates, the symmetry of a specimen is broken in one of paired notches and its initial symmetry loses significance.

2. Low-cycle symmetric three-point bending of a beam with edge notch

In the performed tests, the effect of various loading conditions, change in the geometry of a specimen, and preliminary plastic strain of material from which the specimens were made on the process of damage accumulation were studied. The attempt was made to determine parameters, which may be extended from the particular cases considered in the tests, to more general loading conditions. A possibility for description of regularities of damage accumulations with the aid of simple analytical functions involving constants just as determined from experiments, so specific for every material is considered.

The specimens were made from aluminum alloy D16T in the original state, so from preliminary stretched materials with the various degrees of plastic strain. The composition of the D16T alloy is as follows: Al was as a base metal, the alloying elements were Fe (0.3%), Si (0,19%), Mn (0.76%), Cu (4.0%), Mg (1.29%).

The experiments were conducted on electromechanical testing machine (the rated capacity load was 100 kN). The loading was repeated three-point bending with unloading and it was given by travelling of a moveable cross-head with a constant velocity. Loading diagrams were recorded at each loading cycle.

The minimum force of a cycle was $P_{min} \approx 0$ for all the conducted tests. The maximum force of a cycle P_{max} was considered in three forms: i) $P_{max} = const$ (stationary low-cycle loading), ii) $P_{max} \equiv P_{max}(N)$ was increasing step-function of N for which the number of cycles at one step was constant (non-stationary low-cycle loading with increasing load), and iii) $P_{max} \equiv P_{max}(N)$ was analogous decreasing step-function. The P_{max} value in all tests was chosen to provide plastic material deformation ahead of the notch tip. Tests under stationary cyclic loadings were conducted for different P_{max} values in the range from the limit of elasticity to the strength limit of a specimen. Besides, tests with different notch depth to beam height ratios were conducted.

Increment of the residual deflection $\delta w = \delta w(N)$ depends on the cycle number. The increment of the δw deflection arises mainly due to mechanical properties of material in a pre-fracture zone near the notch tip, and therefore, as it was said before, it is considered as a measure of damage increment in this zone. The increment is considered to achieve the limit w^* value if subsequent deflection of the beam proceeds without increase in P .

2.1 Stationary loading

Fig. 1 demonstrates, as an example, the experimental diagram represented by curves of beam deflection w versus applied force P in stationary low-cycle testing a specimen made from D16T. The P value is sufficiently large in order that fracture to happen after the limited number of cycles (in this case, $N^* = 390$). This allows one to visualize all distinctive features of such diagrams obtained also for test materials for different P_{max} values. Curve 1 corresponds to single loading of a specimen until fracture occurs; group of curves 2 corresponds to cyclic loading up to the instant when a crack starts to extend for $P < P_{max}$. Each curve of group 2 corresponds to loading branch of one cycle. All the curves of group 2, except the first one, have the initial horizontal section $P = 0$. The length of this section is equal to the value of residual deflection accumulated at previous cycles. In this figure, δw is the distance between adjacent curves of group 2.

Damage accumulation can be divided into two specific stages. The first stage (Fig. 1, subgroup A of curves 2) is a stage of cyclic strengthening at which decrease in δw is observed with increase of N. At this stage, δw achieves some minimum that is characteristic for the given value of P_{max} after which the δw value becomes constant within the limits of measurement accuracy. The second stage (Fig.1, sub-group B of curves 2) is characterized by the increase in the δw value as N increases, and this stage is accomplished by growth of a crack when $P < P_{max}$. Accumulation of micro-defects during the first stage is likely to lead to formation of macroscopic defect, which then progresses during the second stage. Therefore, we call the second stage as a stage of development of a macro-defect. The ratio between the number of cycles in sub-groups A and B and the law of δw variation depend on material characteristics (Karpov, 2009; Kornev et al., 2010).

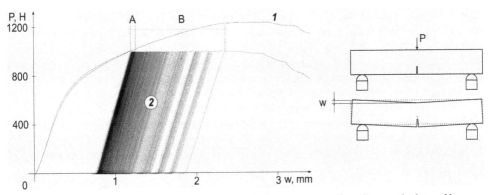

Fig. 1. Scheme of low-cycle test of specimen loaded in tree-point bending and plots of beam deflection as a function of force for every loading cycle for D16T alloy

Regularity in the residual deflection can be visualized as $\delta w(N)$ diagrams. An example of such diagrams is given in Fig. 2. Here pairs of curves are shown with numerals 1-5 for curves, one of the pairs being given by the analytical function and the second one being a saw-like profile. Here the saw-like profiles represent experimental $\delta w(N)$ curves for the D16T alloy. The analysis of experimental diagrams shows that curves can be approximated by plots of some power functions. These functions are to have asymptotes corresponding to the limits beyond which the process described by the diagrams can not take place. That is, the inverse power function with some scaling coefficient can be taken as approximating one.

Besides, in the general case, both descending and ascending branches of the experimental curves should be approximated by different curves.

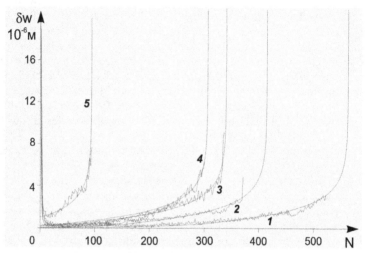

Fig. 2. Approximation of experimental curves of residual deflection increment for D16T alloy

The connection point of these curves corresponds to the minimum δw value. In this case, the form of curve, which approximately describes either descending or ascending branch, is unique for all values of P_{max}. That is, the curves of approximating functions for all P_{max} values can be superposed by parallel shift. Fig. 2. shows such curves (branches of hyperbolas corresponding to saw-like profiles 1-5). These curves are given by functions of the form $f(N) = \lambda (\alpha + \beta N)^{-\gamma}$ normalized in such a way that the experimental minimum point δw would be a connection point of two curves approximating both the descending and ascending branches of the saw-like profile (α, β, γ, λ are experimental constants). If approximating functions have been defined, then the derivatives of these functions calculated for integers of the variable N can be used as a magnitude characterizing material damage. The stage of cyclic strengthening A of duralumin, which differs from steels by essentially larger grain sizes, is limited to several cycles even for low loads (Kornev et al., 2010). As compared with the B stage, the stage A can be neglected.

2.2 Non-stationary loading

Under non-stationary loading conditions, the process of cyclic strengthening begins again each time when P_{max} increases, and only the last stage explicitly includes both A and B stages of damage accumulation. However, if instead of change in δw, the analogous change in $\sum \delta w(N)$ (summation is performed over numbers of all cycles at one loading step), the stages A and B are also evident as in the case of the stationary loading. Fig. 3 demonstrates the experimental diagrams represented by curves of w versus P for non-stationary cyclic loading of alloy D16T.

The non-stationary loading with decreasing P_{max} shows that the initial overload of material provokes consequences, which may affect fracture process when loading with the low P_{max}

value is applied again. Given in Fig. 4 are curves of w versus P for cyclic loading of D16T duralumin when decrease in P_{max} is step-like. Here, after some initial P_{max} value, monotonic nonlinear increase of $\delta w(N)$ is continued even for significant decrease in the P_{max} value. Beginning from the fourth step, $\sum \delta w(N)$ also increases. As a result, in spite of decreasing load, the specimen rapidly losses its load capacity. The tests also show that the increase in P_{max} after several steps of such loading leads to significant inelastic strain with strengthening for values of the force P, which earlier were appropriate to the linear dependence of the deflection on the applied force. This fact evidences that repeated loadings of a material preliminary experienced overloading lead to significant material damage even this loading is not followed by noticeable increase in the residual strain.

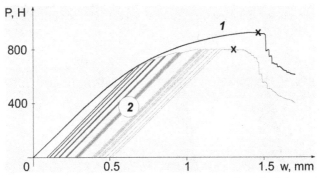

Fig. 3. Non-stationary cyclic loading with increasing maximal applied force for D16T alloy

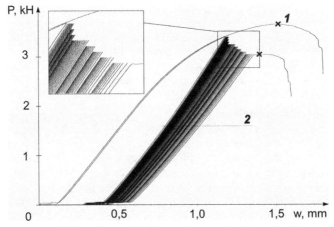

Fig. 4. Non-stationary cyclic loading with decreasing maximal applied force after initial overloading for D16T alloy

2.3 The typical ratio
The typical value for both stationary and non-stationary low-cycle loading is the ratio between limit deflections for single and repeated loading (w^* and w^{**}, respectively). The

tests show that this ratio for material is constant when different schemes of stationary and non-stationary loading are applied (Kornev et al., 2010). This allows one to use this ratio for comparison of results obtained on specimens with various geometrical dimensions and for different loading regimes. For duralumin, we have $w^* \geq w^{**}$.

2.4 Preliminary inelastic strain

Preliminary inelastic strain of a material, from which the specimens have been made, essentially influences material resistance to cyclic fracture. As an example, Fig. 5 displays the experimental diagrams with curves of w versus P (Fig. 5 (a)) and the $\delta w(N)$ diagram (Fig. 5 (b)) for D16T duralumin with various degrees of preliminary stretching: diagram **1** for original materials, diagram **2** for materials stretched by 5%, and diagram **3** for materials stretched by 10%. All the specimens were loaded for the same P_{max} value, but in Fig. 5 (a), diagrams for three tests are displaced, for convenience, from each other along the horizontal axis.

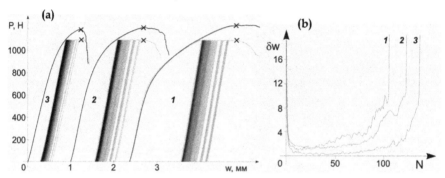

Fig. 5. (a), (b). Low-cycle loading of aluminum alloy after preliminary plastic deformation for D16T alloy

In Fig. 5(b), the area under the curve characterizes the limit deflection w^{**}, which decreases as preliminary stretching increases. However, the decrease in w^{**} is followed by the decrease in δw. This leads to increase of the limiting number of loading cycles.

2.5 Variation in the notch depth

Comparison of tests conducted on beams with notches of different depths has shown that if $\delta w(N)$ diagrams have been plotted for some notch depth l, the diagrams can be used to obtain analogous diagrams for other l values. Assume that $\delta w(N)$ diagrams have been plotted for the notch depth l_1, and for the notch depth l_2 there is the only diagram with curves of w versus P for single loading. The value of maximum applied force $P_{max} = P_1$ corresponds to some deflection $w = w_1$ for l_1 under single loading, the limit deflection of cyclic loading for this notch depth being $w^{**} = w_1^{**}$, and the limit deflection of the single loading being $w^* = w_1^*$. In this case, the curve of damage accumulation $f_1(N)$ for $P_{max} = P_1$, $l = l_1$ can be used to obtain the curve $f_2(N)$ for $P_{max} = P_2$, $l = l_2$. Here P_2 is the force for which a specimen with the notch depth l_2 has the deflection $w = w_2$ such that

$$\frac{w_1}{w_1^{**}} = \frac{w_2}{w_2^{**}} \text{ , where } \frac{w_2^{**}}{w_2^*} = \frac{w_1^{**}}{w_1^*} \Rightarrow w_2^{**} = \frac{w_1^{**} w_2^*}{w_1^*} .$$

3. Development of fatigue crack under low-cycle tension conditions of a plain specimen with edge notch

Tests on a plain specimen loaded in tension with the narrow edge notch were conducted. During the test, direct observation of fatigue crack propagation was performed using a digitized microscope with the resolution of 22500 pixel/mm². The testing scheme is shown in Fig. 6. The field of view is outlined by dashed line.

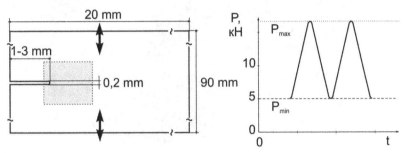

Fig. 6. Test scheme

The specimens were made from D16T alloy preliminary heat-treated at 500°C to give more plastic material. Plain specimens with notches of different lengths (1–3 mm) were used. The minimum load was the same in every cycle, the maximum load was chosen such that three different loading types were provided: *i*) near yield strength, *ii*) near the limit of load capacity, *iii*) the average value of them.

Photographs in Fig. 7 illustrate stages of crack propagation near crack-like defect in two cases *i*) continuous tension with constant rate, *ii*) low-cycle tension.

In the case of continuous tension, the following stages may be observed. First, intense plastic deformation ahead of the notch tip occurs and two zones of strain localization are formed with a delta-shaped area between these zones. The area we term as a pre-fracture zone since further it will define the crack extension direction. At this stage, several focuses of fatigue crack initiation are formed, which are located at notch angles as it is seen in the photograph (a.1). The pre-fracture zone is formed not by a prospective crack, but by the notch itself and its shape is unclear. Further (a.2), one of microcracks develops as a crack propagating within the zone of plastic strain localization irrespective of the pre-fracture zone specified by the notch. This zone becomes more structured and its tip is separated from the crack tip. Here the pre-fracture zone tip starts to shift towards the developing crack. At the next stage (a.3), crack branching takes place, the branches being formed just as near the crack, so at its faces. This evidences the significant extent of material embrittlement in the vicinity of crack extension. The angle at the pre-fracture zone tip starts to decrease. Then (a.4) the branch nearest to the pre-fracture zone tip has some advantages and defines the final direction of crack extension. When the crack tip joins the pre-fracture zone tip (a.5), the critical state is achieved after which the crack starts to extend very fast. The final failure of a specimen is preceded by a short stage (a.6), at which the angle at the pre-fracture zone crack becomes similar to the crack opening angle and one of pre-fracture zone edges defines a path of the subsequent crack extension unambiguously. The crack is very short before the critical state: its length is less than the notch width.

Under repeated low-cycle loading conditions, the pre-fracture zone created by a notch plays no noticeable role especially in the cases when cycle loading starts at insignificant plastic

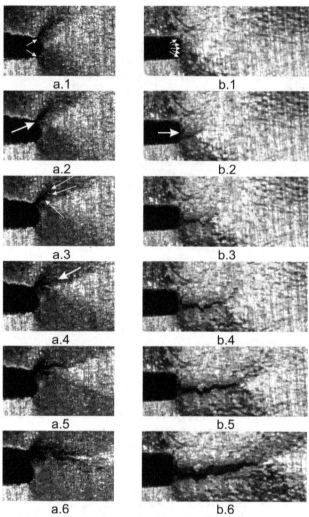

Fig. 7. Stages of crack propagation for D16T alloy; a.1. occurrence of incipient cracks under single loading conditions; b.1. the same under low-cycle loading conditions; a.2. growth of one of incipient cracks under single loading conditions; b.2. the same under low-cycle loading conditions; a.3. crack branching and moving of pre-fracture zone of a notch close together with the crack tip under single loading conditions; b.3. occurrence of "tooth" under low-cycle loading conditions; a.4. development of crack branch nearest to the pre-fracture zone of a notch under single loading conditions; b.4. growth of a crack beyond pre-fracture zone of a notch, formation of the proper pre-fracture zone for a crack under low-cycle loading conditions; a.5. merging pre-fracture zones of the crack and notch under single loading conditions; b.5. final stage of crack development under low-cycle loading conditions characterized by continuous growth of visible sizes of the pre-fracture zone and by decrease of the angle between whiskers; a.6. crack in critical state followed by fast final fracture under single loading conditions; b.6. the same under low-cycle loading conditions

deformations of a specimen. In this case, zones of localization of maximum plastic strains at notch angles are slightly structured and the crack can start to develop from any point of the front bound of the notch (b.1). When one of microcracks starts to extend, a delta-shaped pre-fracture zone is formed ahead of the microcrack (b.2). When the crack is short, the length of the pre-fracture zone is defined by that of the edge notch. The angle between whiskers in the pre-fracture zone tip is close to the right angle, and visible sizes of this zone are of the order of surface roughness occurring due to plastic flow of material. Then, at some distance from the front edge of the notch, temporal crack arrest takes place in all the considered cases. Keeping constant length, the crack begins to open at the expense of blunting and increasing the length of its front edge, and then it changes both the direction and rate of propagation. The path, which the crack takes, deviates from the mean direction of its propagation and the former coincides with the upper or lower edge of the pre-fracture zone. After this, the crack returns to the mean path of propagation. The "tooth" formed at a crack face is clearly seen (b.3). Then the rate of crack growth is steadily increased, crack opening being continued. The pre-fracture zone length increases (b.4). The next to last stage is characterized by continuous growth of visible sizes of the pre-fracture zone and by decrease of the angle between whiskers (b.5). At last, the critical state is achieved (b.6): the crack is blunted, significant crack opening takes place that is comparable with the original notch width, then the crack produces a path branch sharply deviated from the mean path of propagation and then final failure of material occurs after which loading must be ceased. As opposed to continuous loading, the crack length to the instant of the critical state is close to the notch length.

Given in Fig. 8 are plots characterizing crack propagation as a function of the number of cycles. At the left, some geometric characteristics of cracks versus the number of cycles are given. At the right, $(P-\varepsilon)$ plots of specimens are given, where P is applied force, ε is the overall elongation of a specimen. Both minimum and maximum loads of loading cycles are shown on these plots, as well as the maximum strain on the first loading cycle. The value of maximum force in a cycle was chosen such that to provide areas of plastic strain ahead the notch tip.

Plots a.1 and a.2, b.1. and b.2, and c.1. and c.2 correspond to specimens with the notch of 1 mm, 2 mm and 3 mm in length, respectively. Curves 1, 2, 3 and 4 on the left plots correspond to the overall length of crack along its face, the distance of the crack tip from the front notch edge, the difference between these values, and the notch width increase of which shows opening of the crack mouth, respectively. All these values demonstrate nonlinear growth of cracks for which hyperbolic functions are applicable for description of both this growth and damage accumulation at the notch tip in the case of three-point bending of a beam. The general structure of material undoubtedly influences crack propagation as it can be seen on plots in Fig. 8 (a.1, b.1, and c.1), deviations of a crack due to structural heterogeneity being leveled out at its further propagation.

Fig. 9. illustrates separate parts of a crack corresponding to different stages of cyclic loading. Here a crack at the notch 3 mm in length is shown.

It seems likely that origination of a "tooth" is not a consequence of a heterogeneous structure of the material, but this is associated with sizes of defects from which a fatigue cracks start to propagate. In this case, a residual durability of the structure can be defined by its appearance.

Let us discuss interaction between plasticity zones formed in the vicinity of the notch tip and propagating cracks under single and cyclic loading conditions. Fig. 11 shows: a) schemes of zones of plasticity localization in the vicinity of the notch tip (region I) with features near angles (regions II and III), b) schemes of crack initiation under single loading conditions, c) scheme of crack initiation under cyclic loading conditions, d) scheme of fatigue crack propagation with origination of a "tooth" at the boundary of the zone of plasticity localization for a notch. Only in the case when load capacity is exhausted, initiation of cracks on the front notch edge at several points takes place: in original material under single loading conditions (Fig. 11.b), in material accumulated damages under fatigue loading (Fig. 11.c).

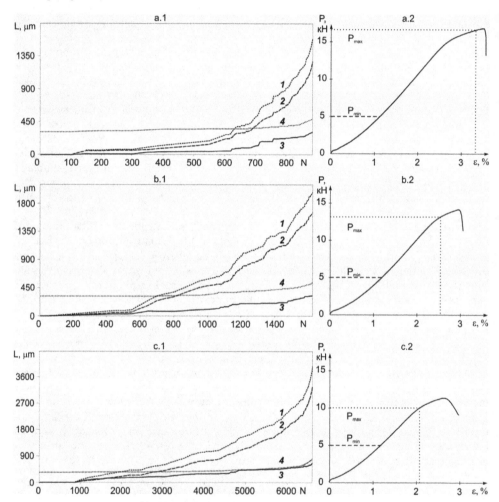

Fig. 8. Characteristics of fatigue crack propagation for D16T alloy

Given in Fig. 10 are forms of three cracks for notches of various lengths (notch lengths in mm are shown inside contours of final pre-fracture zones). Here it is seen that a "tooth" is inherent to all the cracks, the distances from the crack onsets to "teeth" are in direct proportion to lengths of notches. This "tooth" can be identified with pronounced fatigue striations originated on crack faces. In Fig. 10, all three "teeth" are marked with circles. The distance to a "tooth" is close to one third of the overall length of each crack. The segment on the horizontal axis in Fig. 8 is assigned to origination of a "tooth". This segment is nearly the same for all the cracks with respect to the total number of cycles. These segments correspond to about 600, 1100 and 4800 cycles, respectively, on a.1, b.1, c.1.

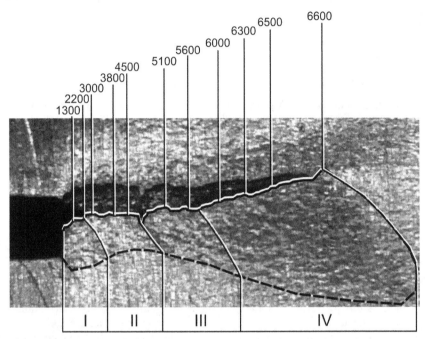

Fig. 9. Propagation of crack in pre-fracture zone for D16T alloy. Given at the top are the numbers of loading cycles, dashed line at the bottom outlines the visible area of maximum plastic strains. This area is divided into four regions: a crack initiates and propagates from the beginning in the closed state, and then it takes the mean propagation path (region I); the stage of stable crack propagation before a "tooth" arises (region II); stable crack propagation after "tooth" proceeds with increasing rate and continuous growth of the pre-fracture zone (region III); the final stage of crack propagations that is characterized on plots by clear nonlinearity (region IV)

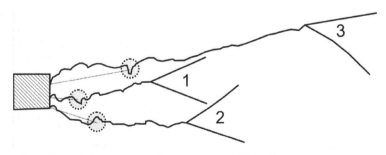

Fig. 10. Cracks and pre-fracture zones for various notch lengths

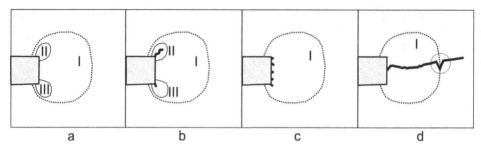

a   b   c   d

Fig. 11. Cracks and pre-fracture zones

At the initial stage, a fatigue crack passes through material located in the region of plasticity localization near the notch (region I in Fig 11.a, regions I and II in Fig. 9). Further, when the crack continues to propagate, the zone of plasticity localization ahead of the sharp crack tip serves as a pre-fracture zone. The former zone is sufficiently small. As a result of the fact that the crack tip passes from one region to another, a pronounced fatigue striation is originated in the form of a "tooth" (marked by circle in Fig 11.d).

4. Comparison between test results and theoretical notions

The model of low-cycle fatigue describing pulsing loading of a specimen with the inner macrocrack has been proposed in (Kornev, 2004, 2010). This model is appropriate to the scheme by Laird-Smith (Laird & Smith, 1962; Laird, 1967). Within the framework of the proposed model in (Kornev, 2004, 2010), information on material strain in the pre-fracture zone has been obtained in detail: processes of damage accumulation, step-wise crack tip advance, and failure of structures for pulsing loading are described. Accumulation of damages is associated with inelastic strain of materials in the pre-fracture zone. The simple relations for the critical fracture parameters and the fatigue life have been obtained. Attention should be paid to the following circumstances: in the foregoing model, the information of damage accumulation and the hypothesis concerning the crack arrest are essentially used. Emphasize that when damages are accumulated, just as linear, so nonlinear summation of damage in the pre-fracture zone material may occur in the context of the considered model. In deciding between one and another way for summation of damages, no experimental data on damage accumulation at every loading step were available. The experimental data described in the previous section make up this deficiency.

The model in (Kornev, 2004, 2010) describes occurrence of striations under fatigue fracture. Material is considered to consist of quasi-brittle fibers separated by thin layers, which possess quasi-ductile fracture type before strain, and after inelastic strain of layers, the fracture type changes to quasi-brittle. The fiber diameters coincide with diameters of grains of tested materials (the diameters of grains are $\approx 10^{-2}$ cm), and the fiber widths coincide with the thickness of layers separating subgrains. The width of the layers is $\approx 10^{-4}$ cm. Further we assign numeral 1 to the fiber material and numeral 2 to the thin layer material. Properties of the layer material allow description of occurrence of some marks (whiskers or ears) from fatigue striations. In essence, in work (Kornev, 2004, 2010) there is considered the behavior of the simplest composite medium, material of which changes its fracture type under inelastic strain.

For deriving sufficient fracture criteria (Kornev, 2004, 2010) for low-cycle fatigue, modification of the classical Leonov-Panasyk-Dugdale model (Kornev, 2004) is used, where the pre-fracture zone is a rectangle ahead of the crack tip. The modification of the classical Leonov-Panasyk-Dugdale model allowed one to describe not only the pre-fracture zone length $\Delta_{1\sigma+}$ at every loading cycle, but a magnitude of inelastic strain under tension $\varepsilon_{\sigma+}$ for material of the pre-fracture zone fiber nearest to the macrocrack center

$$\varepsilon_{\sigma+} = \frac{1 - \dfrac{\sigma_{m1}}{\sigma_a}\sqrt{\dfrac{r_1}{2l}}\dfrac{k_1}{\sqrt{n_1}}}{\dfrac{5}{\pi(\eta+1)}\dfrac{G_1}{\sigma_{m1}}},$$

$$\frac{\Delta_{1\sigma+}}{r_1} = \frac{k_1^2}{2n_1}\frac{\left(\dfrac{5\sqrt{2}}{4(\eta+1)}\dfrac{G_1}{\sigma_{m1}}\varepsilon_{\sigma+}\right)^2}{\left(1 - \dfrac{5}{\pi(\eta+1)}\dfrac{G_1}{\sigma_{m1}}\varepsilon_{\sigma+}\right)^2} \tag{1}$$

Here σ_a is the amplitude of pulse loading; σ_{m1} is the limit of elasticity; n_1 and k_1 are integers ($n_1 \geq k_1$, k_1 is the number of damage-free material fibers); $n_1 r_1$ is the averaging interval for the first material; r_1 is the specific linear dimension of the first material structure; $2l_0$ and $2l = 2l_0 + 2\Delta_{1\sigma+}$ are lengths of initial and fictitious cracks, respectively; G_1 is the shear modulus of fibers; $\eta = 3 - 4\mu$ and $\eta = (3-\mu)/(1+\mu)$ are coefficients for plane strain and plane stress state, respectively, where μ is the Poisson ratio; for relations (1), the restriction $1 - 5G\varepsilon_{\sigma+}/(\pi\sigma_{m1}(\eta+1)) > 0$ holds.

Under cyclic pulse loading conditions, when the scheme of three-point bending is used, hysteresis loops take the form given in Fig. 12.

These hysteresis loops with translation differ from the standard statement in the model in (Kornev, 2004, 2010), in which the scheme of rigid loading under unloading is accepted. In Fig. 12, $\varepsilon_1^{(i)}$ is the limit elongation of original materials for $i = 0$; that after the first inelastic strain is $i = 1$; that after the second inelastic strain is $i = 2$, and etc. First three loops are depicted with lines widen from one loop to another, and the onset of the fourth loop is depicted with dots.

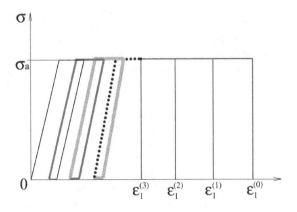

Fig. 12. Scheme of material damage

Consider a typical situation. Let failure occur at the fourth loading cycle. During the strain process, the material elongations at the real crack tip coincide with the limit material elongation $\varepsilon_1^{(3)}$ after the third inelastic strain of the fiber nearest to the real crack tip (Fig. 12), and the crack tip being advanced.

Recall that the crack length in the model in (Kornev, 2004, 2010) changes by the pre-fracture zone length $\Delta_{1\sigma+}$ after the step-wise real crack tip advance, and under the repeated loading, materials in the pre-fracture zone become brittle (Romaniv et al., 1990; Laird & Smith, 1962; Kornev, 2004, 2010). Each such advance is associated with the certain number of cycles when linear and non-linear damages are summed. The performed tests show that the initial state of material, which falls into the pre-fracture zone, influences the process of the step-wise crack tip advance. The magnitude of inelastic strain under stretching $\varepsilon_{\sigma+}$ and the pre-fracture zone length $\Delta_{1\sigma+}$ depend on: i) load amplitude σ_a ; ii) initial crack length $2l$; and iii) preliminary inelastic strain of material. If the basic parameter of inelastic strain $\varepsilon_{\sigma+}$ slightly depends on $2l$, then for the pre-fracture length $\Delta_{1\sigma+}$, the analogous dependence is pronounced. This corresponds to the passage of the second fracture process to the third one on the Paris curve (Romaniv et al., 1990; Shaniavski, 2003).

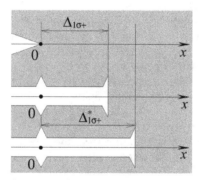

Fig. 13. Schematic drawing of material damage

As explained the step-wise advance of the real crack tip by the length $\Delta_{1\sigma+}^* = \chi\Delta_{1\sigma+}$, where χ is such a coefficient that for quasi-ductile layers $\chi = 1$, and for quasi-brittle layers $\chi > 1$.

The most important question is as follows: what is a distance from the crack onset where the crack will arrest? In the model in (Kornev, 2004, 2010), the case of original material loading is considered when the layers possess the quasi-ductile fracture type. Because of this, the ratio $\chi = 1$ holds. After preliminary inelastic strain of a material from which specimens have been made, embrittlement of layer materials occurs, and the ratio $\chi > 1$ can hold. The step-wise crack tip advance in composite materials with quasi-ductile and embrittled layers elucidates the schematic drawing in Fig. 13. At the scheme top, the right-hand tip of a blunted crack is shown before its start. In the middle part of the scheme, the crack tip advances by the segment $\Delta_{1\sigma+}$, then the crack is blunted at the mesoscale, see the ratio $\chi = 1$. At the scheme bottom, the crack tip advances by the segment $\chi\Delta_{1\sigma+}$, and then the crack is blunted at the mesoscale, see the ratio $\chi > 1$.

In essence, just as in the model in (Kornev, 2004, 2010), so in foregoing experimental results, the distinct influence of the single parameter on the process of damage accumulation under cyclic pulse loading is traced. In the model in (Kornev, 2004, 2010), it is the parameter that characterizes inelastic strain of material in the pre-fracture zone $\varepsilon_{\sigma+}$, and in experiments, it is the parameter, describing inelastic (residual) deflection of beams. In experiments, the residual deflection of beams characterizes both advance of the real crack tip and development of the pre-fracture zone after each loading cycle. The experimental results obtained agree with the diagram of fatigue failure with various kinds of relief elements formed as a result of fatigue fracture seen in Fig. 3.23 from (Shaniavski, 2003), just as the theory described in (Kornev, 2004, 2010), so the experimental data being correlated with the second and third stages of the Paris curve.

The model in (Kornev, 2004, 2010) gives no preference to linear or nonlinear summation of damages in metals (Romaniv et al., 1990; Coffin & Schenectady, 1954) under cyclic loading. The performed tests show that damage accumulation, in general, is always nonlinear, however, if the loading is performed for P_{max} near to the limit of the specimen plasticity, there exists, as a rule, the large interval on the axis of the number of cycles N, where the δw value is close to its minimum for given P_{max} value. Within this interval, the damage accumulation may be considered to be linear as it is under non-stationary loading with P_{max} values close to the limit of plasticity. In both cases, the model proposed in (Kornev, 2004, 2010) and the obtained experimental results agree well. In this case, the hypothesis concerning a crack arrest holds (Kornev, 2004, 2010): a crack arrests at the interface between a fiber and a layer and then it is blunted. This layer is the first one located beyond a pre-fracture zone $\Delta_{1\sigma+}$, and the zone itself is located in material not subjected to some plastic strain.

When a loading regime with increased P_{max} is considered (this regime models catastrophic overload of a structure with crack), the model like that in (Kornev, 2004, 2010) is proposed, and the experimental results obtained agree essentially worse. Such disagreement of theoretical notions (Kornev, 2004, 2010) with the experimental results is explained by the relations $\chi = 1$ or $\chi > 1$. After the first catastrophic series of loadings, the intensity of damage accumulation process in material significantly increases. In this loading regime, the refined hypothesis $\chi > 1$ $\chi > 1$ concerning crack arrest holds: a crack arrests at the fiber-layer interface at the distance $\Delta_{1\sigma+}^{*} > \Delta_{1\sigma+}$ and it is blunted, this interlayer is located beyond the pre-fracture zone $\Delta_{1\sigma+}$, and the last is in material that has already been subjected to some plastic strain and embrittlement (Romaniv et al., 1990; Laird & Smith, 1962; Kornev, 2004, 2010).

5. Conclusion

Recording the loading diagrams for beams with the edge crack allows description of both damage accumulation at the macroscale and failure of constructions at every loading cycle. The damage accumulation in the pre-fracture zone is associated with the residual deflection of beams after unloading.

Preliminary plastic deformation of aluminum alloy leads to decrease in its durability. However, here the intensity of damage accumulation also reduces under low-cycle three-point bending of beams. This results in increase in the number of loads, which structure may withstand.

In the chapter, loading regimes with gradual overloading and with increased loading have been considered. These regimes model a common situation and catastrophic overloading under cyclic loading. The results obtained may be useful for prediction of fatigue life of a structure with the crack and in analyzing situations after overloading.

Direct observation in the vicinity of the notch tip during the process of low-cycle tension of a plain specimen allows one to trace the behavior of a fatigue crack at various stages of its propagation: *i*) crack initiation in the vicinity of stress concentration; *ii*) crack propagation within the plasticity zone; *iii*) onset of generation of a narrow pre-fracture zone formed by the crack itself; *iiii*) fast crack propagation beyond the plasticity zone.

Schemes allowing description of deformation, damage accumulation, and failure of material under fatigue with account of the preliminary inelastic deformation of the material and effect of stress concentration on crack initiation have been proposed.

6. Acknowledgment

The work was financially supported by Russian Foundation for Basic research (No 08-01-00220), in the context of the project No 23.16 included into the program of Presidium of Russian Academy of Sciences, and of *Integration Project of SB RAS, UB RAS, FEB RAS No 119*.

7. References

Romaniv, O.N.; Yarema, S.Ya.; Nikiforchin, G.N.; Makhutov, N.A.; Stadnik, M.M. (1990). *Fatigue and cyclic fracture toughness of structural materials, Vol. 4. Fracture mechanics and strength of materials, in four volumes*, Naukova Dumka, ISBN 5-12-000489-X, Kiev, Russia (in Russian)

Shaniavski, A.A. (2003). *Safety fatigue fracture of elements of aircraft constructions. Synergetic in engineering applications*. Monografiya, ISBN 5-94920-015-2, Ufa, Russia (in Russian)

Laird, C.; Smith, G.C. Crack propagation in high stress fatigue. *The Philosophical Magazine, A. Journal of Theor. Experim. and Aplied Physics*, Vol.7. No.77, (1962), pp. 847-857, ISSN 1478-6435

Laird, C. The influence of metallurgical structure on the mechanism of fatigure crack propagation. *Fatigue Crack Propagation, ASTM STP 415; Am. Soc. Testing Mats.* (1967), pp. 131-168, ISBN 0-8031-1250-5

Kornev, V.M. Two-scale model of low-cycle fatigue. Change from quasi-ductile to brittle fracture. *Strain and fracture of materials*. No.2, (2008), pp. 2-11 (in Russian), ISSN 1814-4632

Kornev, V.M. Distribution of stresses and crack opening displacement in the pre-fracture zone (Neuber-Novozhilov approach). *Physical Mesomechanics*. Vol.7, No.32, (2004), pp. 53-62 (in Russian), ISSN 1683-805X

Nikitenko, A.F. (1997). *Yield and long strength of metallurgical materials*. Institute of Hydrodynamics, Siberian Branch of Russian Academy of Sciences, Novosibirsk, Russia (in Russian), ISBN 5-7795-0024-X

Coffin, L.F.; Schenectady, N.Y. A Study of the effects of cyclic thermal stresses on a ductile metal. *Transactions of the ASME*. Vol.76., No.6., (1954), pp. 931-950, ISSN 0742-4795

Karpov, E.V. Deformation and fracture of a spheroplast under low-cycle loading at various temperatures. *Journal of Applied Mechanics and Technical Physics*, Vol.50, No.1, (2009), pp. 163–169, ISSN 0869-5032

Kornev, V.M. Two-scale model of low-cycle fatigue. Embrittlement of pre-fracture zone material. *Procedia Engineering*. Vol.2, No.1, (2010), pp. 453-463, ISSN 1877-7058

Kornev, V.; Karpov, E.; Demeshkin, A. Damage accumulation in the pre-fracture zone under lowcyclic loading of specimens with the edge crack. *Procedia Engineering*. Vol.2, No.1, (2010), pp. 465-474, ISSN 1877-7058

Part 2

Microstructures, Nanostructures and Image Analysis

6

Microstructural Evolution During the Homogenization of Al-Zn-Mg Aluminum Alloys

Ali Reza Eivani[1,2], Jie Zhou[2] and Jurek Duszczyk[2]
[1]Materials Innovation Institute (M2i), Mekelweg 2, 2628 CD Delft,
[2]Department of Materials Science and Engineering, Delft University of Technology,
Mekelweg 2, 2628 CD Delft,
The Netherlands

1. Introduction

1.1 Background

Aluminum and aluminum alloys are probably the most ideal materials for extrusion, and they are the most commonly extruded. Most of commercially available aluminum alloys can be extruded. Principal applications include parts for the aircraft and aerospace industries, pipes, wires, rods, bars, tubes, hollow shapes, cable sheathing, for the building, automotive and electrical industries. Sections can be extruded from heat-treatable or non-heat treatable low-, medium- and high-strength aluminum alloys [1].

In the last 30 years, the development of aluminum extrusion technology has, in the main, been focused on the billet metallurgy, die design and process control for low- and medium-strength aluminum alloys in the 6xxx series for architectural applications, in order to maximize extrusion speed and at the same time fulfill the requirements in product specifications in terms of dimensions, shape, surface and mechanical properties. As a result, there is a wealth of information available on the relationship between alloy chemistry, microstructure and extrudability of these alloys [2]. In comparison, the fundamental knowledge and extrusion technology, especially those for medium- and high-strength aluminum alloys in 7xxx series, are rather scarce in the open literature [2].

7xxx series aluminum alloys, almost exclusively for air transport applications in the past but now increasingly used in the rail and road vehicles, must comply with much more stringent performance specifications than 6xxx series aluminum alloys for architectural applications. Although many investigations on the behavior of medium- and high-strength aluminum alloys at individual processing steps have been performed, systematic research linking all these processing steps is lacking, while the extrusion behavior is associated with alloy composition and a series of microstructural evolutions throughout the whole chain of material processing from casting through homogenization to extrusion. Such research is particularly needed for the aluminum extrusion companies that are currently shifting the application fields of extrusions from architecture to ground transport where medium-strength alloys (7003, 7005, 7010, 7020, 2011, 2017 and 2618) and high-strength alloys (7049, 7050, 7075 and 2024) are increasingly used. This chapter concerns one of the mostly used medium-strength alloys, AA7020, as a representative of Cu-free 7xxx series aluminum alloys. Table 1 shows the nominal chemical composition of the AA7020 aluminum alloy.

Element	Si	Fe	Cu	Mn	Mg	Zn	Ti	Cr	Zr	Al
Wt. %	<0.35	<0.35	<0.2	0.05 – 0.5	1.0 – 1.4	4.0 - 5.0	Zr + Ti = 0.08-0.25	0.1 – 0.4	0.08 – 0.20	Bal.

Table 1. Nominal chemical composition of the AA7020 aluminum alloy

It should be noted that despite a broad range of applications, the AA7xxx series alloys have a number of characteristics that are not favorable for material processing, for example, low extrudability, high extrusion pressure required together with low solidus temperature which can cause incipient melting, makes their production at low throughputs. In addition, in the 7xxx series aluminum alloys including the AA7020 alloy, due to the long solidification interval (the temperature gap between the liquidus and solidus), microsegregation tends to be significant and homogenization needs lots of time, which causes the formation of second phase particles, some of which may be detrimental to the hot workability of the alloy as well as its final mechanical properties. Furthermore, the formation of a peripheral coarse grain structure is quite common in these alloys which can significantly degrade the mechanical properties. These issues will be discussed as limiting factors in the extrusion of the AA7020 aluminum alloy.

1.2 Limiting factors in the extrusion of AA7020

Numerous limiting factors, e.g., the formation of defects, low extrudability and the complications in the final microstructure of the product in 7xxx series aluminum alloys can impair the productivity of the extrusion process [2]. Most of the limiting factors are directly or indirectly related to the chemical composition, metallurgical features formed DC casting and evolving during the homogenization treatment, and extrusion conditions [2]. Using an optimum chemical composition within the allowance range of a specific alloy, in combination with optimum homogenization and extrusion conditions can result in a significant increase of the extrudability of the material and improvement of the mechanical properties of the final product. This requires the knowledge of the factors presented below, i.e., hot workability and peripheral recrystallization, both of which are strongly dependent on chemical composition of the alloy and the homogenization treatment.

1.2.1 Hot workability

For wrought aluminum alloys, hot workability is an important index of manufacturability. It refers to the capacity of an alloy to withstand hot deformation at a maximum rate without inducing flow non-uniformity or structural defects. If translated into extrusion (termed extrudability), it is defined as the maximum speed for a sound extrudate with sufficient dimension and shape accuracy. Hot workability is in fact affected by all parameters affecting the fracture of the material under processing [1]. It is strongly dependent on the size and density of second-phase particles which are in turn dependent on the chemical composition and homogenization treatment. On the other hand, compressive stresses superimposed on shear stresses during the deformation process can have a significant influence on closing small cavities or limiting their growth and thus enhancing workability. Because of the important role of the stress state, it is not possible to express workability in absolute terms. Workability depends not only on material characteristics but also on process variables, such as strain, strain rate, temperature, and stress state [1]. In other words, extrudability can be deteriorated by two factors: (i) unfavorable processing conditions and (ii) the presence of large second-phase particles.

Investigation of the effect of processing conditions on the hot workability of AA7xxx series aluminum alloys is out of the scope of this chapter. However, the effect of particles which would deteriorate the hot workability of the alloy is the main subject of this chapter.

High-strength aluminum alloys (7xxx series) are known for having rather poor hot workability due to the presence of dissolved and later precipitated elements in the form of large particles that raise flow stress and cause the actual temperature of the extrudate to increase above the solidus temperature, leading to hot tearing as shown in Fig. 1. Hot tearing represents the separation or failure of the product as a result of a sequence of phenomena consisting of local melting, crack formation and final fracture of the product. Hot tearing may occur as a result of the presence of large intermetallic particles or even the multiplication of the two mentioned factors (local melting and large particles). Therefore, in hot extrusion, applicable extrusion speed depends on the alloy composition and the microstructure formed during direct-chill casting and developed during homogenization, cooling and reheating to the initial billet temperature.

Fig. 1. Extreme case of hot tearing in AA7075 aluminum alloy [3]

In the case of the 7xxx series alloys, the two phases that deteriorate the extrudability are (i) compounds of Al-Fe-Mn-Si elements, which are especially important if they are located at the grain boundaries and (ii) Al-Mg-Zn-Cu eutectic phases which are mostly located at the dendrite boundaries.

1.2.1.1 Particles composed of Al, Fe, Mn and Si

It has been proven that in the case of the AA7020 aluminum alloy Al, Fe, Mn and Si-containing particles have the $Al_{17}(Fe_{3.2},Mn_{0.8})Si_2$ stoichiometric composition and are mostly located at the grain boundaries and therefore are called GB particles [4]. GB particles deteriorate the hot workability of the alloy in addition to mechanical properties since they are hard and brittle especially when located at the grain boundary regions. In order to avoid the detrimental effects of these particles, they should be dissolved during the homogenization treatment. If the particles are not dissolvable, they should be spheroidized.

1.2.1.2 Particles composed of Al, Mg, Zn and Cu

During the solidification of the 7xxx series aluminum alloys containing Mg, Zn and Cu, some intermetallic particles such as Al_6CuMg_4, $Al_2Mg_3Zn_3$, $AlCuMg$, $MgZn_2$, Al_2Cu and $MgZn_2$ phases form [5-9]. In addition, mutual solutions of different phases can result in the formation of new particles, for example, mutual solid solution of Al_6CuMg_4 and $Al_2Mg_3Zn_3$ compounds (T phase), solid solution between $AlCuMg$ and $MgZn_2$ compounds (M phase), solid solution formed by $Al_5Cu_6Mg_2$ and Mg_2Zn compounds (Z phase) and between Al_2CuMg and Al_2Cu compounds (S phase). The investigation of these particles is interesting for hot working since they mostly have low melting points, which may result in incipient melting during hot deformation.

1.2.2 Peripheral recrystallization

In addition to high flow stress and low solidus temperature, local recrystallization is another factor limiting the extrusion window of medium- and high-strength aluminum alloys. For ground and air transport applications, a qualified extrudate should not contain any undesirable microstructural features, most notably local recrystallized, excessively large grains.

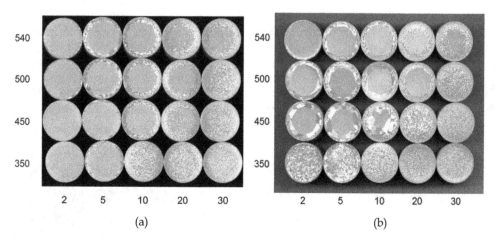

Fig. 2. Macrostructures of AA6005 25 mm bar varying with ram speed (mm/s) and billet temperature (a) as extruded and (b) after solution heat treatment [10, 11]

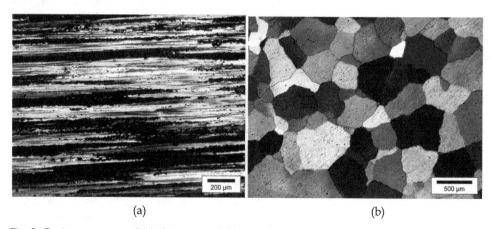

Fig. 3. Grain structures of (a) the core and (b) periphery of AA7020 extruded at 450 °C, 3 m/min and 15:1 [12]

The microstructure of an extruded aluminum product varies from an unrecrystallized fibrous structure to a thoroughly recrystallized fine or coarse grain structure (Fig. 2), depending on the chemical composition, homogenization treatment and extrusion conditions, i.e., speed and temperature, and cooling procedure [10, 11]. The most undesirable

microstructure is the one with a peripheral coarse grain (PCG) structure [2] which is a well known defect in hot extruded aluminum alloys. In this case, the peripheral surfaces of the structure are fully recrystallized, having large grains, while the core is composed of unrecrystallized elongated grains as shown in Fig. 3.
PCG degrades the properties of the extruded product such as strength, fracture toughness and stress corrosion resistance [2]. It is actually a perpetual problem that extruders encounter in meeting the specifications of aircraft alloys that base their strength requirements on typical longitudinal properties of the unrecrystallized core and assume implicitly that no recrystallized outer band structure is present. It is generally understood that the peripheral recrystallization is a complex interplay of billet composition (grain growth inhibitors, i.e., Mn, Cr, or Zr), microstructure, deformation conditions and critical temperatures (solvus, solidus and recrystallization) [2, 3, 10-12].

1.3 Application of homogenization treatment

A homogenization treatment after DC casting for the 7xxx alloys is meant to serve the purposes of dissolving second-phase particles and generating disposoids that are able to inhibit recrystallization and PCG zone formation. The metallurgical features that occur during DC casting and should be studied during the homogenization treatment are presented below.
1. The mechanical properties of extruded products are largely dependent on alloying elements present in solid solution. These elements increase the strength mainly through solid solution or precipitation hardening [13]. During casting of aluminum alloys, a large fraction of alloying elements segregate to the liquid and result in an inhomogeneous distribution of alloying elements. Therefore, removal of the inhomogeneous distribution of alloying elements on a microscale is of prime importance during the homogenization treatment.
2. Segregation can also result in the formation of eutectic constitutive particles with low melting points in the grain boundary regions or inside the grains [14, 15]. As mentioned earlier, the presence of low melting point (LMP) phases which may cause incipient melting during hot deformation can deteriorate the hot workability of aluminum alloys. Therefore, one of the aims of the homogenization treatment is to dissolve LMP phases.
3. In addition, the formation of some hard particles with sharp edges mostly from impurities, e.g., Fe, in combination with some alloying elements such as Mn and Si is expected during DC casting. These particles also decrease the hot workability and limit the range of process parameters applicable during extrusion [2, 14, 16-19]. Therefore, it is necessary to dissolve these particles as well, in order to obtain high mechanical properties and extrudability. If the dissolution of these particles is exhibitively energy and time consuming, these particles should be spheroidized.
4. Eliminating the PCG structure or decreasing its extent is of great interest to the aluminum extrusion industry. It is generally known that the formation of small dispersoid particles can pin the low and high angle grain boundaries and therefore, inhibit recrystallization and grain growth. Therefore, an optimum homogenization treatment should take the formation of fine, well-distributed dispersoid particles into account.

1.4 Previous works on homogenization treatment of aluminum alloys

Although there have been a number of investigations on the homogenization treatment of the 7xxx series aluminum alloys in recent years [5-8, 20-26], most of the efforts have been focused on the nature and evolution of the $Al_2Mg_3Zn_3$ (T), Al_2CuMg (S), $(CuZnAl)_2Mg$ and $MgZn_2$ (η) phases [5-9] and the formation and distribution of dispersoids during homogenization [20-26]. In addition, in comparison with other aluminum alloys, the information on the 7xxx series aluminum alloys is rather scarce in the literature. Although some researchers have studied the microstructural changes and the evolution of the eutectic and low melting point phases during homogenization in the case of the 7xxx series aluminum alloys, there is still a lack of information in the case of the AA7020 aluminum alloy. Moreover, there has been no comprehensive quantitative study on the dependence of the particles on homogenization treatment parameters. Most of the investigations are concerned with the microstructural evolution and phase transformations during homogenization, describing the phenomena in a qualitative manner. For example, Lim et. al. [6] investigated the effects of compositional changes and preheating conditions on the evolution of constitutive particles, the M, T, S phases and dispersoids in AA7175 and AA7050 alloys. Senkov et. al. [27] studied the effect of homogenization treatment on the microstructural evolution of four newly developed 7xxx series aluminum alloys to obtain optimized conditions. Jackson and Sheppard [14] studied the effect of homogenization treatment on the microstructural changes of AA7075, 7150 and 7049 alloys. They focused on the evolution of the microstructure, low melting point phases and the M, T, S, Al_3Mg_2 (b), β, $Al_{18}Cr_2Mg$ (E) and $CrAl_7$ phases. Fan et al. [28] studied the evolution of microstructure in an Al-Zn-Mg-Cu alloy during homogenization. Ciach et. al. [29] conducted theoretical and experimental studies on the dendritic structure and its dissolution in aluminum-zinc alloys. However, the research on the commercially important AA7020 alloy is scarce. No quantitative investigation on the microstructural evolution in AA7020 regarding the grain boundary particles and low melting point phases during homogenization has been carried out.

As mentioned earlier, in addition to the dissolution of particles, dispersoid formation is another important aim of the homogenization treatment. Dispersoids exert a retarding force or pressure on low angle and high angle grain boundaries, which has a profound effect on the behavior of aluminum alloys in terms of recovery, recrystallization and grain growth [30]. A large amount of experimental research in addition to modeling efforts [15, 20, 24-26, 31-42] has been carried out to understand the conditions under which the dispersoids form and the effect of homogenization parameters on the evolution of dispersoids, with the aim of maximizing the effect of pinning the grain boundaries to retard recrystallization and grain growth. Most of the previous research [15, 20, 25, 42] focused on Al_3Zr as the only dispersoids present in the microstructures of the AA7xxx series aluminum alloys. However, the formation of the other types of dispersoids in other series of aluminum alloys has been a subject of extensive research. For example, in the AA3xxx series aluminum alloys, the formation of Mn-containing dispersoids and their effect on the recrystallization behavior were investigated [31-33]. In the AA6xxx series aluminum alloys, different kinds of dispersoids, i.e., Zr-, Mn- and Cr-containing ones were found to play individual roles in recrystallization inhibition [34-38]. In the case of the AA7xxx series aluminum alloys, the formation of Zr- and Sc-containing dispersoids has been investigated [15, 20, 25, 42]. For example, Robson et al. [15, 20, 25, 42], investigated the effect of Zr addition on the dispersoid formation and recrystallization fraction after hot deformation. It was concluded that using an optimum two-step homogenization treatment, a smaller fraction of recrystallization could be obtained. Robson [25] further studied the effect of Sc on the formation of dispersoids, as Sc was expected to eliminate the dispersoid free zones, as observed in the scandium free AA7050 alloy, thus greatly increasing the recrystallization resistance.

Comprehensive investigations to characterize Cr- and Mn-containing dispersoids formed in the AA7xxx series aluminum alloys during homogenization are scarce, although Cr-, Zr-, and Mn-containing dispersoids commonly co-exist in these alloys. Since analyzing all the individual dispersoids during the investigations to optimize the homogenization treatments is practically impossible, clear characterization of different types of dispersoids with a combination of analytical methods is of prime importance.

The objectives of this research were to determine the effect of homogenization treatment on the evolution of the particles, especially the grain boundary and low melting point ones, and to establish the correlations of the process parameters such as time and temperature with the fractions of these particles in the structure. Thorough investigations were performed on the effect of homogenization treatment on the evolution of these particles using optical microscopy (OM), X-ray diffraction (XRD) analysis, field emission gun-scanning electron microscopy (FEG-SEM), electron probe microanalysis (EPMA) and differential scanning calorimetry (DSC) and the results were quantitatively analyzed. The dependence of the fractions of the particles in the structure on homogenization parameters was also investigated using various quantitative methods. In addition, a comprehensive investigation on the formation of Cr-, Zr-, and Mn-containing dispersoids which commonly co-exist in the AA7020 aluminum alloy was performed.

2. Experimental procedure and data processing

Cubic samples of 2 cm in size were cut from the centre of a direct-chill (DC) cast AA7020 ingot. The chemical compositions of the variants of the AA7020 alloy used in this study are shown in Table 2. Isothermal homogenization treatments were performed in a salt bath at temperatures of 390-550 °C for 2-48 h, as shown in Table 3. Following the heat treatments, the samples were quenched in water.

Element (Wt %)	Si	Fe	Cu	Mn	Mg	Zn	Ti	Cr	Zr	Al
N1	0.31	0.28	0.2	0.34	1.24	4.36	0.001	0.10	0.08	Bal.
N2	0.30	0.30	0.19	0.35	1.20	4.37	0.002	0.10	0.13	Bal.
N3	0.29	0.31	0.2	0.36	1.22	4.37	0.001	0.10	0.20	Bal.
Nominal composition	<0.35	<0.35	<0.2	0.05-0.5	1.0-1.4	4.0-5.0	Zr+Ti= 0.08-0.25	0.1-0.4	0.08-0.20	Bal.

Table 2. Chemical compositions of the AA7020 alloy variants used in this study

Temperature (°C)	Time (hours)
	2
390	4
430	8
470	16
510	24
550	32
	48

Table 3. Homogenization treatment conditions used in this study

Optical microscopy (OM) was performed using an OLYMPUS BX60M light microscope on the samples etched using Barker's etchant. Images were analyzed using the Soft Imaging Software (SIS) image processor. Three samples in each homogenization condition were prepared and the analysis was performed on two images with approximately 6.2 mega pixel image quality and the average values are reported. The differences between the measured data from different samples and different images are represented by error bars.

The samples were examined using field emission gun-scanning electron microscope (FEG-SEM). The optimum operating voltage and current were 10 kV and 1 nA, respectively. With these settings, dispersoids as small as 10 nm in diameter could be detected.

The SEM images of the GB particles after different homogenization treatments were quantitatively analyzed to investigate their dissolution during homogenization. 20 GB particles were analyzed in each case, and the width was measured and the average value calculated. During homogenization at high temperatures, i.e., 510 and 550 °C, some of the GB particles were completely dissolved in the structure. The dissolved GB particles were also considered in the calculation with a null width. The average initial number density of the GB particles in 20 micrographs of the structure was counted to be 2×10^9 μm^{-2}. The average number density of the GB particles after homogenization was also counted employing the same method and, if it was less than the average initial number density, indicating the full dissolution of some of the GB particles, a zero width was put into the calculations.

Discs having a diameter of 3 mm were punched from the samples and ground down to a thickness of less than 60 μm, followed by electro-polishing in a solution of 30% nitric acid and 70% methanol cooled to -25 °C in a double-jet polishing unit at 20V.

Energy dispersive X-ray (EDX) analysis was performed with an analyzer attached to the FEG-SEM to determine the chemical compositions of the particles in the as-homogenized microstructures. In the case of small particles (< 500 nm), in order to keep the analysis volume in the EDX measurements as small as possible, the analysis was performed on TEM samples with an average thickness of 100 nm or less.

Electron Probe Microanalysis (EPMA) was performed using an electron beam with energy of 15 keV and beam current of 50 nA employing Wavelength Dispersive Spectrometry (WDS). The composition at each analysis location of the sample was determined using the X-ray intensities of the constituent elements after background correction relative to the corresponding intensities of reference materials. The thus obtained intensity ratios were processed with a matrix correction program CITZAF [43]. The points of analysis were located on lines with increments of 2 μm and involved the elements of Cr, Mn, Cu and Zr. Al was measured by difference.

A BRUKER-AXS D5005 diffractometer with Cu Kα1 wavelength was used to identify the phases present in the as-cast and as-homogenized conditions. Quantitative XRD (QXRD) analysis was performed using the direct comparison method [44] to estimate the weight percent of the phases in the structure. Application of this method requires the weight percent of the phase of interest (i.e., GB particles) in the as-cast structure, as the baseline. To calculate the weight percent of the GB particles in the as-cast structure, the surface fraction of the GB particles was calculated using FEG-SEM together with EDX analysis. The analysis was performed on 20 images at a magnification of 1000 and all the particles present in each image were analyzed. Assuming a uniform distribution of the GB particles in the structure, the surface fraction can be approximated to be equal to the volume fraction. The volume fraction of the GB particles was converted to weight percent using the density of the GB particles (3709 kg/m³ [45, 46]) and the density of AA7020 aluminum alloy (2780 kg/m³ [47]). The only assumption made was the density of the other particles (a mixture of various phases) other than the GB particles being equal to the density of the AA7020 aluminum alloy.

To determine the volume fraction of particles from the data obtained by optical and SEM microscopy, a simple rule was used. It was assumed that the average surface fraction measured in a large number of images from different positions in the substrate was representative of the volume fraction [48]. It has been mathematically proven that the average surface fraction is equal to the volume fraction, provided that an enough large number of sections are investigated [48]. In this research, the investigation was performed on such a number of images that a constant average value was obtained, being not significantly changed by adding another image to the measurements.

The number density and radii of dispersoids obtained from SEM micrographs are in the form of the number of particles on many cross sections in the observation area in 2-D. 2-D cross section observations of the volume generally do not directly correspond to the coherent values in 3-D. In other words, the average particle diameter and number density of particles in each size group of the size distribution are not correct representatives of the real numbers in 3-D. The reason is that the crossing plane may not cut the particle in the middle and therefore, an observed specific cross section with a constant size may be a cross section of a particle which is cut through the middle or a cross section of a larger particle which is not cut through the middle. This point is schematically illustrated in Fig. 4 [48]. It can be seen that a mono-dispersed system of diameter D_j in 3-D can result in different circular sections in 2-D. It is shown in Fig. 5 that particles of large sizes can contribute to increasing the 2-D observed number density of particles with smaller sizes depending on the geometry of the cutting plane.

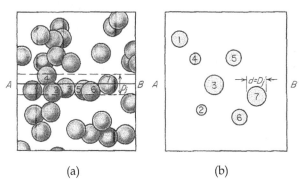

(a)   (b)

Fig. 4. (a) Distribution of mono-sized particles of D_j in 3-D and (b) observed A-B cross section of the particle distribution in 3-D [48]

The solution to this problem is to subtract the contribution of large particles to the 2-D measured sizes of smaller particles. For this purpose, different methods such as Scheil's method, Schwartz's method, Schwartz-Saltykov method have been proposed and used [48]. These methods can be used to find the distribution of particle sizes from a distribution of section diameters. The three methods differ in the details of how the numbers of sections contributed by larger spheres are determined.

In addition to the methods mentioned above, there are other methods which work with the distribution of section areas to determine 3-D particle sizes. Among these methods Johnson's and Johnson-Saltikov methods are well known. Johnson's derivation is applicable only to single-phase structures. However, Saltykov's improvement of Johnson's method applies to a distribution of particles as well as grain sizes [48]. Since the method is applicable for the prediction of grain and subgrain sizes in addition to particles, Johnson-Saltykov method was used in this research.

According to Saltykov's method [48], the most rational scale for the classification of particles (or grain sizes) is a linear logarithmic scale of diameter. Using the Johnson-Saltykov method, the analysis and calculations in the logarithmic scale can be simplified and facilitated. An advantage of this method is that a size distribution of particles can be obtained directly [48]. However, it must be noted that the resulting size distribution graphs will be presented by logarithmic size categories.

Thermal analysis of the as-cast and homogenized materials was carried out by means of a DSC analyzer at a heating rate of 20 °C/min over a temperature range of 35 to 700 °C. Samples were cubes weighing 12 mg each and Al_2O_3 powder was used as the reference. The analysis was performed under the protection of Ar gas. To ascertain the effect of homogenization treatment on the dissolution of the LMP phases the DSC profiles were quantitatively processed. For this purpose, the area underneath the peak was correlated to the fraction of the LMP phases in the structure [3].

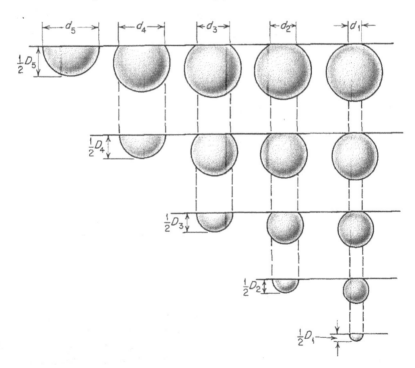

Fig. 5. Contribution of single size particles, i.e., D_5 in different 2-D size groups depending on the geometry of cutting plane [48]

3. The as-cast microstructure

3.1 Grain boundary (GB) particles

Low and higher magnification secondary electron FEG-SEM images of the as-cast microstructure of an AA7020 aluminum alloy variant (N2) are shown in Fig. 6. The constitutive particles elongated along the grain boundaries can be clearly seen. The average width of these grain boundary (GB) particles is 640 nm. The perturbations on the surfaces of the GB particles are illustrated by arrows in Fig. 6. (b).

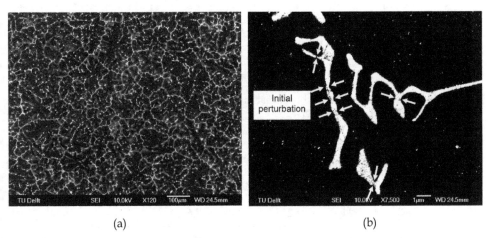

(a)   (b)

Fig. 6. (a) Low and (b) higher magnification SEM micrographs showing the GB particles in the as-cast microstructure (alloy variant N2) [49]

To determine the compositions of the GB particles, EDX analysis on more than 20 GB particles having the same morphology was performed. The results showed that the majority of the GB particles had similar compositions, as given in Table 4. By using an image analyzing software together with EDX analysis on a large number of different secondary phases in the as-cast structure, the initial fraction of the GB particles with respect to all of the secondary phases was calculated to be 74±3 wt.%.

With XRD analysis, a phase in a mixture can be identified if its volume fraction is higher than 5% [44]. The results of the image analysis indicated that the volume fraction of the GB particles was close to 7%. Therefore, it was possible to determine the identity of these particles using XRD analysis [44]. The results, shown in Fig. 7, illustrate that only one secondary phase could be detected, which was $Al_{17}(Fe_{3.2},Mn_{0.8})Si_2$ (PDF No. 01-071-4015 [45]). Comparison of the XRD results with the EDX analysis, as given in Table 4, shows a good agreement.

The chemical composition of thermodynamically stable Al-Fe-Mn-Si compounds may be presented by $Al_{16}(Fe,Mn)_4Si_3$ or $Al_{15}(Fe,Mn)_3Si_2$ [4]. The crystallography of the intermetallic phases containing aluminum, silicon, iron and manganese implies that they should be considered as phases with multiple sublattices [50]. Therefore, these compounds may be simply considered as a solution of the Al-Fe-Si particles and Mn or vice versa. In this case, their formation and stability at different conditions obey the thermodynamics of solutions. Since Fe and Mn can reside on the same sublattices [50], the Al-Fe-Mn-Si particles can be considered $Al_{16}(Fe_{(1-y)},Mn_y)_4Si_3$. From the role of the $RT((1-y)\ln(1-y)+y\ln y)$ term in the Gibbs free energy of solutions [51], it can be stated that a compound with equal values of Fe and Mn has the lowest total free energy.

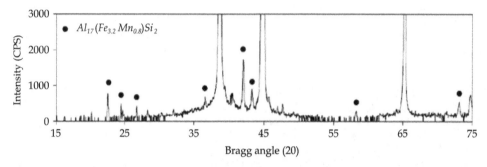

Fig. 7. X-ray diffraction pattern of the as-cast material showing the presence of the GB particles in the alloy variant N2 [49]

Element	Al	Fe	Mg	Si	Zn	Cu
EDX	72.1	16.1	2.8	4.3	2.7	2.0
XRD	62.2	24.2	6.0	7.6		

Table 4. Measured mean compositions (wt. %) of the grain boundary constitutive particles in the as-cast material (alloy variant N2) together with the calculated chemical compositions of the suggested phase identity based on the XRD analysis

In the DC-cast AA7020 aluminum alloy, the amount of Fe is larger than Mn in the grain boundary regions. The larger amount of Fe may be attributed to the partitioning coefficients of Fe and Mn, which result in severer microsegregation of Fe toward the grain boundaries and therefore, a higher concentration of Fe in these regions [52]. Fe has a small solid solubility in aluminum [52]. Therefore, the excess Fe rather than what is consumed in Al-Fe-Mn-Si particles must form other intermetallic compounds. In this case, if y=0.5, in addition to the thermodynamically stable $Al_{16}(Fe_{(1-y)},Mn_y)_4Si_3$ phase, some separate Al-Fe-Si and Al-Fe particles are expected to form to consume the remaining insoluble Fe at the grain boundaries. However, as mentioned above, the solution formation results in a decrease in the Gibbs free energy of the system determined by the $-RT((1-y)\ln(1-y)+y\ln y)$ term. Therefore, in this system, the stable Al-Fe-Mn-Si particles dissolve some of the excess Fe and form the meta-stable $Al_{17}(Fe_{3.2},Mn_{0.8})Si_2$ particles and the remaining Fe incorporates in other intermetallic compounds. The amount of the Fe dissolved in the stoichiometric $Al_{16}(Fe_{(1-y)},Mn_y)_4Si_3$ particles should be so much that the total energy of the system is minimized by the formation of $Al_{17}(Fe_{3.2},Mn_{0.8})Si_2$, Al-Fe-Si and Al-Fe particles. The same may be valid for the replacement of Si atom with exceeding Al in the compound from the stoichiometric values.

3.2 Low melting point (LMP) phases
The presence of the low melting point (LMP) phases in the as-cast structure was determined using DSC. The DSC profile of the as-cast structure is shown in Fig. 8. It is clear that there is

an endothermic reaction at 576 °C and the melting of the bulk sample occurs at 661 °C. In order to ensure that the endothermic peak is associated with the melting of the LMP phases rather than their dissolution, two samples were heated at 10 °C higher and lower than the reaction temperature, i.e., 566 and 586 °C, for 1 min. DSC analysis of these samples showed that, after these treatments, the endothermic reaction still occurred, which indicated that the 1 min treatment did not result in the dissolution of the corresponding phases.

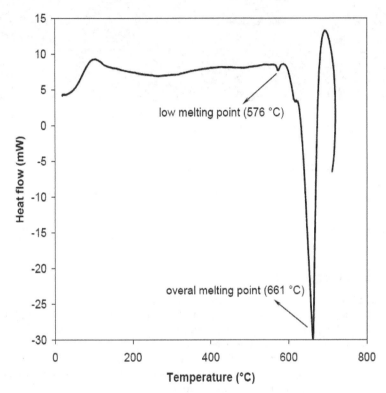

Fig. 8. DCS profile of the as-cast sample (alloy variant N2) at a heating rate of 20 °C/min [53]

The microstructures of the samples were investigated using field emission gun-scanning electron microscope (FEG-SEM). During the analysis, the phases in the as-cast structure, for example, $Al_{17}(Fe_{3.2}Mn_{0.8})Si_2$ and Al-Fe-Si, were detected whose compositions and morphologies were the same as those present in the as-cast structure. The only difference observed in the structures was that for the sample treated at 586 °C for 1 min, the morphology of the Al-Cu-Mg-Zn particles changed from a round shape in the as-cast structure, shown in Fig. 9 (a), to a sponge-like one with perturbations as shown in Fig. 9 (b). These morphological changes must be due to the melting of Al-Cu-Mg-Zn particles during heating up to 586 °C and re-solidification of the particles upon water quenching. This suggests that the endothermic reaction observed in the DSC profile is indeed due to the melting of Al-Cu-Mg-Zn particles. The primary elements present in the LMP phases and their concentrations are shown in Table 5, based on the EDX analysis.

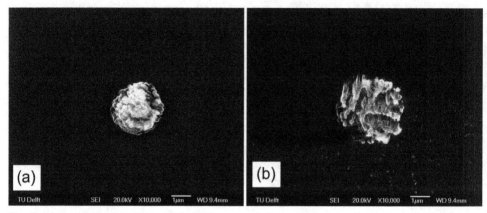

Fig. 9. (a) An Al-Cu-Mg-Zn particle with a round shape in the as-cast structure and (b) a sponge-like Al-Cu-Mg-Zn particle with perturbations in sample N2 after heating to 586°C for 1 min and water quenching [53]

Element	Al	Cu	Mg	Zn
wt%	57±7	17±3	8±2	6±2

Table 5. Measured mean composition of the Al-Cu-Mg-Zn particles in the as-cast alloy variant N2

4. Effects of homogenization

4.1 Microstructural evolution during homogenization

Fig. 10 shows the optical microstructures of the material (alloy variant N2) after 2 h homogenization at different temperatures. Homogenization at 390 and 430 °C led to an increase in the volume fraction of particles. At 470 °C, the volume fraction appeared to be unchanged, while at 510 and 550 °C, it decreased.

Fig. 11 shows low and higher magnification secondary electron FEG-SEM images of the dominant particles formed during homogenization at 390 °C. The grain boundaries are still delineated by the GB particles while the initial continuity of the GB particles shown in Fig. 6(a), is deteriorated by spheroidization. Moreover, large needle-shaped and round precipitates appear in the structure. Examples of these precipitates together with large Al-Fe-Si particles are illustrated in Fig. 11(b). These particles, as pointed at in Fig. 11(a), are dispersed inside the grains. EDX analysis on more than 20 particles with similar morphologies determined the chemical compositions of these precipitates and the results are shown in Table 6.

It was possible to identify these compounds formed during homogenization at low temperatures using XRD analysis. The results given in Fig. 12 (a) show that in addition to the previously present GB particles (Fig. 7), new particle are present in the homogenized microstructure, i.e., $MgZn_2$ (η) and Mg_2Si (β) particles. However, the XRD pattern of the sample homogenized at 550 °C, presented in Fig. 12 (b), shows that no new particles have been formed during homogenization at such a high temperature, which is consistent with the results from the optical microscopy analysis, shown in Fig. 10.

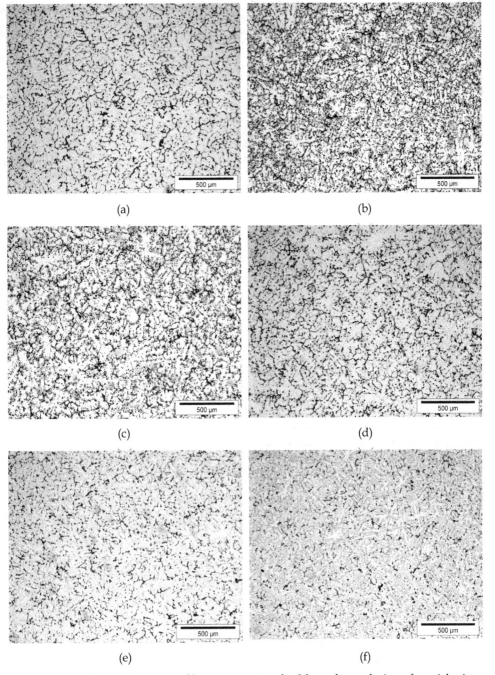

(a)   (b)

(c)   (d)

(e)   (f)

Fig. 10. Effect of the temperature of homogenization for 2 h on the evolution of particles in alloy variant N2, (a) the initial structure, (b) 390, (c) 430, (d) 470, (e) 510 and (f) 550 °C [49]

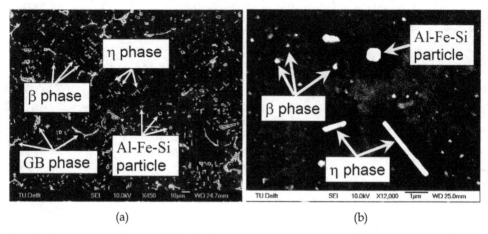

(a)   (b)

Fig. 11. (a) Low magnification FEG-SEM image of the alloy variant N2 homogenized at 390 °C, showing the GB particles and (b) the needle-shaped and round $MgZn_2$ (η) and Mg_2Si (β) particles together with large Al-Fe-Si particles [49]

Element (Wt%)	Al	Mg	Zn	Si	Fe
η phase (EDX)	63±4	4±2	28±1	3±1	2±1
η phase (XRD)	...	15.7	84.3	...	...
β phase (EDX)	56±4	22±3	5±2	15±3	2±2
β phase (XRD)	...	63.38	...	36.62	...

Table 6. Measured mean compositions (wt. %) of the needle-shaped and round precipitates in the as-homogenized microstructure of the alloy variant N2 together with the stoichiometric chemical compositions based on the XRD results

It was also found that even after homogenization at a high temperature, i.e., 550 °C, some of the particles were not dissolved in the structure. These retained particles are mostly the GB particles and other particles which together with their EDX spectrums are shown in Fig. 13 (a) and (b). EDX suggested that the particles shown in Fig. 13 (a) and (b) were $Al_{13}Fe_4$ and Al_8Fe_2Si, respectively.

The investigations carried out using the FEG-SEM of the samples homogenized at 390 and 430 °C indicated the presence of needle-shaped and round precipitates, as shown in Fig. 11. The morphologies of these particles and their chemical compositions indicated these particles to be $MgZn_2$ and Mg_2Si precipitates, which is in agreement with [54-56]. The formation of these precipitates may be attributed to the super-saturation of the structure with alloying elements occurring during solidification at high cooling rates applied during DC casting. When the as-cast alloy is exposed to a homogenization treatment at a low temperature (< 470 °C), there is a tendency for the alloying elements to precipitate out. As the temperature increases (> 470 °C), the solubilities of these elements in the α-Al matrix increase [4, 57] and the formation of new particles is not expected. Thus, it can be concluded that the formation of new particles or the dissolution of old ones depend primarily on the homogenization temperature.

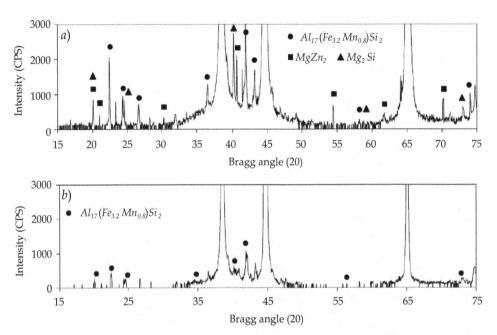

Fig. 12. X-ray diffraction patterns of the alloy variant N2 homogenized at (a) 430 and (b) 550 °C showing the presence of the GB particles, $MgZn_2$ (η) and Mg_2Si (β) particles [49]

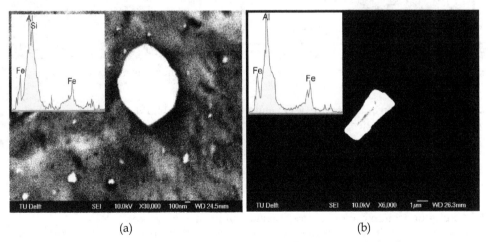

(a)   (b)

Fig. 13. Particles remaining in the microstructure of the alloy variant N2 after homogenization at 550 °C for 48 h, (a) Al_8Fe_2Si and (b) $Al_{13}Fe_4$ particle [49]

4.2 Evolution of the GB particles during homogenization

The evolution of a typical GB particle during homogenization at 390 and 550 °C is shown in Figs. 14 and 15, respectively. It is clear that the dominant process at lower temperatures is the spheroidization of the GB particle, while at higher temperatures the decrease in the width of the GB particle is the main evolution process.

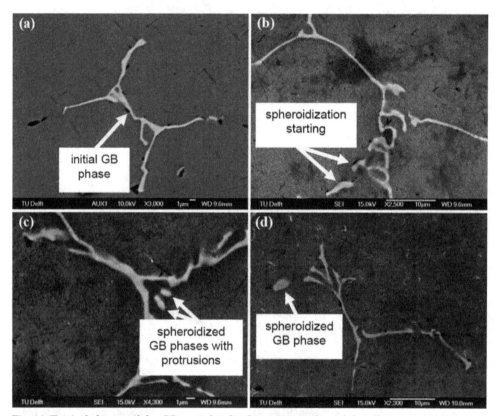

Fig. 14. Typical shapes of the GB particle after homogenization of the alloy variant N2 at 390 °C, (a) initial, (b) 2, (c) 8 and (d) 24 h [49]

Unlike the evolution of other particles in aluminum alloys [2, 14, 15], the evolution of the GB particles during homogenization, depending on the process parameters and the nature of the particles, may occur in the form of spheroidization or dissolution. The spheroidization mechanism of these particles is quite interesting. However, more interestingly, the dissolution of the GB particles obeys a specific dissolution mechanism introduced hereafter as the thinning, discontinuation and full dissolution (TDFD) mechanism.

4.2.1 Spheroidization during homogenization at low temperatures

The analysis of the SEM images indicates that although the fraction of the GB particles does not decrease during homogenization at 390 °C, the morphological changes towards spheroidization take place, as can be seen in Fig. 14. The analysis of 20 pictures from the as-cast structure and the one homogenized at 390 °C for 48 h indicates that after homogenization the fraction of spheroidized particles increases by two times compared with the as-cast structure.

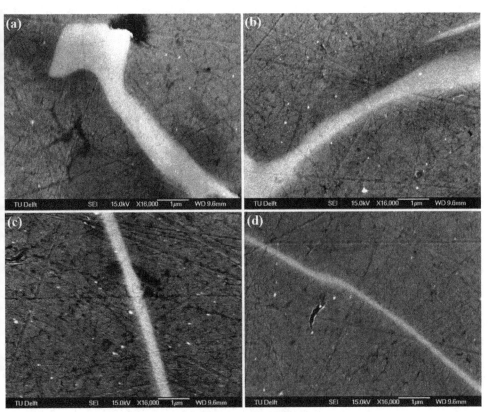

Fig. 15. Decrease in the width of a GB particle after homogenization of the alloy variant N2 at 550 °C, (a) initial, (b) 2, (c) 8 and (d) 24 h [49]

The proposed mechanism of the spheroidization of the GB particles is illustrated in Fig. 16, based on the experimental observations from the FEG-SEM images (a typical one is shown in Fig. 14). Fig. 16(a) shows a GB particle with initial protrusions on its surface. Afterwards, spheroidization occurs and the GB particle takes an ellipse shape, Fig. 16(b). The spheroidization continues till the GB particle takes a spherical shape with protrusions on its surface, Fig 16(c), and the process ends with removing the protrusions till the GB particle resembles a sphere, Fig. 16(d). The driving force for spheroidization is the decrease in the surface energy of the GB particle with decreasing interfacial length between the GB particle and the aluminum matrix [15, 58].

As mentioned earlier, one of the main aims of homogenization treatment prior to hot deformation is to dissolve detrimental particles, especially those located at the grain boundary regions. Although this goal would not be achieved if the particles are not dissolved but spheroidized, spheroidization of particles can be beneficial in the sense that these particles cause less stress concentrations at sharp tips and edges and therefore, inhibit crack initiation which can lead to an improvement in the hot workability of the material.

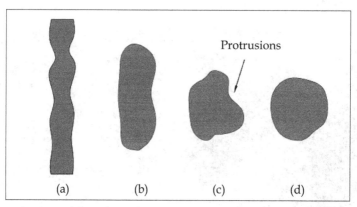

Fig. 16. Schematic view of the spheroidization mechanism describing the evolution of a GB particle during homogenization [49]

4.2.2 Thinning, discontinuation and full dissolution (TDFD) during homogenization at high temperatures

In order to understand the dissolution sequence of the GB particles at high homogenization temperatures, the evolution of a GB particle was investigated at different time intervals during homogenization at 550 °C. It was found that the dissolution process started with the thinning of the GB particle without primary spheroidization. Fig. 15 shows that the average width of the GB particle decreases from 640 nm to 130 nm by a homogenization treatment at 550 °C for 24 h. The thinning process continues until the GB particle become discontinuous in some regions (Fig. 17) and finally the full dissolution of the GB particles occurs. The occurrence of discontinuities during spheroidization of an eutectic particle has been reported elsewhere, for example in [59].

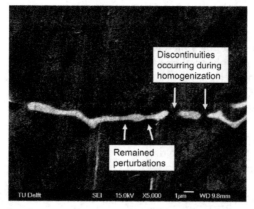

Fig. 17. A GB particle after homogenization at 550 °C for 8 h, showing the thinning and discontinuation (alloy variant N2)

The dissolution mechanism of a GB particle at different stages is schematically illustrated in Fig. 18. The driving force is the increases in the solubility limits in the matrix at high temperatures and therefore the presence of concentration gradients of Mn, Fe and Si in the structure. Fig. 18(a-c) schematically illustrates the overall thinning process of a GB particle. Assuming that during homogenization an overall decrease in the width of a GB particle occurs at a constant rate in different regions regardless of the widths, Fig. 18(a) through (c), the parts having smaller widths meet each other sooner than other parts, as shown by arrows in Fig. 18(d). Therefore, the discontinuities, Fig. 18(e), occur as a result of the inherent perturbations, Fig. 6(b), of the surfaces of the GB particle, shown by arrows in Fig. 18(a). Afterwards, the dissolution continues with the same mechanism as occurring to the small parts till the GB particle disappears. The remaining perturbations which help the continuation of dissolution of the GB particle with a similar mechanism are shown by arrows.

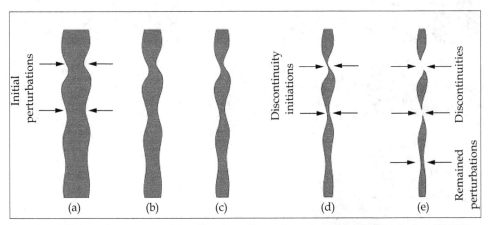

Fig. 18. Schematic description of the thinning, discontinuation and full dissolution (TDFD) mechanism responsible for the evolution of a GB particle during homogenization at high temperatures [49]

The experimental observations of the shapes of the particles, presented in Fig. 19, show that the tips of the particles may initially have rectangular, ellipsoidal, needle or circular cross sections, Fig. 19(a) through (c). However, as shown in Fig. 19(d) through (f), during the dissolution, the tips get a circular cross section. Assume the initial shape of the tips to be rectangular having two steep tips at the edges. According to the Gibbs-Thompson equation [15], a large concentration of the solute at the interface is resulted in, which indeed increases the dissolution rate significantly. The steep edges dissolve sooner and therefore a circular tip will be produced. The same is valid for an ellipsoidal cross section.

5. Quantitative analysis of particle dissolution

5.1 All the particles
5.1.1 Results of quantitative optical microscopy (QOM)
Fig. 20 shows the effect of homogenization time on the volume fraction of all the particles present in the structure. The calculation was based on the changes in the volume fraction of the particles in the structure, as shown in Fig. 10. It is clear that during homogenization at

390 and 430 °C the fraction of particles increases, while during homogenization at 510 and 550 °C the fraction of the particles decreases, indicating dissolution.

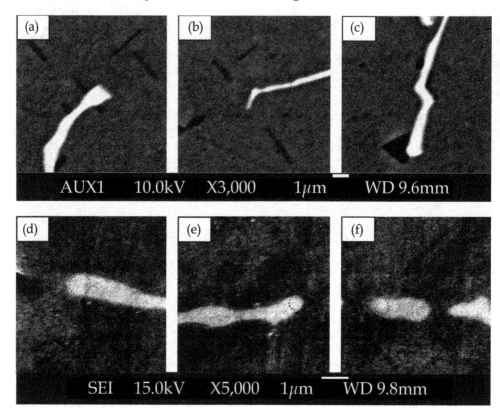

Fig. 19. Shapes of the tips of an $Al_{17}(Fe_{3.2},Mn_{0.8})Si_2$ particle in the alloy variant N2, (a) through (c) in the initial structures and (d) through (f) after homogenization at 550 °C for 8 h. In (d) through (f), circles have been drawn on the tips of the particles, showing the perfectness of the circular cross section of the tips [60]

The most noticeable in Fig. 20 is the increase in the fraction of all the particles at the initial stage of homogenization at low temperatures, which acts in contrary to the aim of the homogenization process [2, 14, 15]. The increases in the fraction of particles during the first 2 h of homogenization at 390 and 430 °C, as shown in Fig. 20, are due to the formation of a large number of new particles, namely $MgZn_2$ (η) and Mg_2Si (β). As stated above, it indicates that part of the elements precipitate out by forming precipitates during the first 2 h of homogenization.

5.2 GB particles
5.2.1 Results of quantitative XRD analysis (QXRD)
Fig. 21 compares the strongest XRD peaks and other ones related to the GB particles in the AA7020 alloy variant N2 samples homogenized at different temperatures for 8 h. This figure was obtained by focusing on the 41 to 44 ° Bragg's angle (2θ) range in the XRD patterns, i.e.,

Figs. 7 and 12. Compared with the as-cast structure, it is obvious that the intensity of the peaks related to the GB particles in the sample homogenized at 390 °C is almost unchanged, while at 430 and 470 °C it is decreased slightly. Homogenization treatment at higher temperatures, namely 510 and 550 °C, however, resulted in marked decreases in the intensity of the peaks related to the GB particles in the XRD pattern in comparison with that of the as-cast structure.

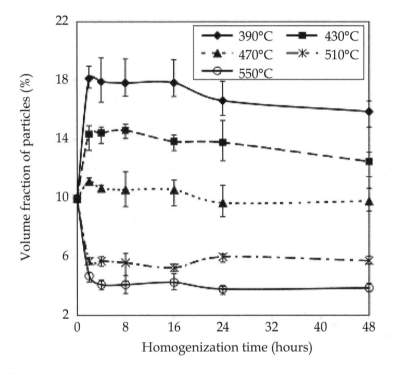

Fig. 20. Effect of homogenization time on the volume fraction of particles (alloy variant N2) [49]

To quantify the results, the fraction of the GB particles in the structure after homogenization was calculated according to the results of the XRD analysis using the direct comparison method [44]. As it was mentioned earlier, the initial fraction of the GB particles with respect to all of the secondary phases in the as-cast structure, necessary to form a baseline for the direct comparison method, was calculated to be 74±3 wt.%. Fig. 22 illustrates the weight percents of the GB particles as a function of homogenization time at different temperatures. It can be seen that the fraction of the GB particles after homogenization at 390 °C remains unchanged and at 430 °C it is decreased slightly, which is not in agreement with the results of the quantitative optical microscopy (QOM) shown in Fig. 20. Therefore, the increase in the fraction of particles during homogenization shown in Fig. 20 is due to the formation of new precipitates (η and β), but not due to the increase in the GB particles. However, at higher temperatures, the fraction of the GB particles decreases, which is in line with the behavior shown in the optical microscopy measurements.

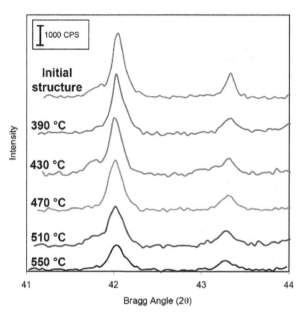

Fig. 21. The strongest (left) and another (right) XRD peaks related to the GB particles in samples homogenized at different temperatures for 8 h (alloy variant N2) [49]

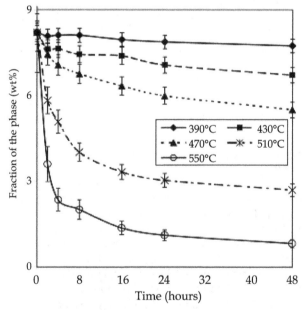

Fig. 22. Fraction of the GB particles (wt.%) in the structure of the alloy variant N2 after homogenization at various temperatures, based on the QXRD analysis [49]

5.2.2 Results of quantitative analysis using FEG-SEM (QSEM)

The measured average widths of the GB particles as a function of homogenization time are shown in Fig. 23. It can be seen that, at 390 and 430 °C, the average width of the GB particles is almost unchanged from the initial value. The minor variations in the width of the GB particles at these temperatures are within the margin of error. However, at higher temperatures, dissolution occurs, evidenced by the decreases in the width of the GB particles. It can also be seen in Fig. 23 that the most rapid dissolution occurs at the first few hours of homogenization regardless of homogenization temperature. Moreover, the dissolution rate is much higher at 550 °C.

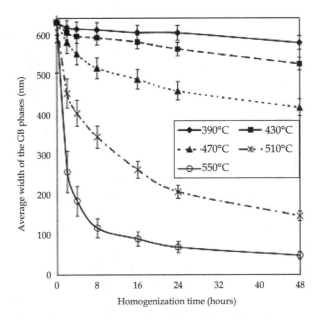

Fig. 23. Average GB widths in the alloy variant N2 as a function of homogenization time, based on the QSEM analysis [49]

To account for the dissolution of the GB particles, a study on the solubility limits of the elements composing the main GB particles, namely Fe, Mn and Si, is essential. The data on the solubility limits of the elements in Al-Mn-Fe-Si are scarce in the literature [4]. In this case, the solubilities of the elements in Al in the four-component Al-Mn-Fe-Si regions adjacent to the ternary systems may be estimated, based on three-component regions of the Al-Fe-Mn, Al-Fe-Si and Al-Mn-Si ternary systems [4]. Table 7 shows the solid solubility limits of iron, manganese and silicon in the four component Al-Fe-Mn-Si system at different temperatures [4]. It is clear that at low temperatures, i.e., lower than 470 °C, the solubility limits of the elements forming the GB particles are small in the α-Al matrix. Therefore, considerable dissolution of the GB particles is not expected at these temperatures. This is in agreement with the results of the quantitative image analysis of the OM images, presented in Figs. 20, 21 and 22. At higher temperatures, however, the solubility limits increase, as shown in Table 7, which indicates that the most of the alloying elements in the GB particles are dissolved in the α-Al matrix. It is also clear that even at such high temperatures (e.g., 550

°C), the solubility limits of the elements in the Al matrix are less than the weight percentages of those elements in the composition of the alloy (N2). Therefore, complete dissolution of all the GB particles is not expected even after holding for a long time. This explains the observations in Fig. 10 (f) and 15 (d) that there are still some GB particles remaining in the structure after homogenization at 550 °C. EDX analysis confirmed the existence of these particles even after homogenization at 550 °C for 48 h.

	Solubility limit Wt%		
T (°C)	Fe	Mn	Si
390	0.002	0.026	0.03
430	0.004	0.05	0.06
470	0.009	0.15	0.08
510	0.016	0.25	0.11
550	0.044	0.44	0.2

Table 7. Solid solubility limits of iron, manganese and silicon in the four component Al-Fe-Mn-Si system at different temperatures [4]

5.2.3 Comparison between the QXRD and QSEM results
The benefit of using QXRD analysis is that the results obtained are primarily related to the GB particles while the QOM analysis includes the GB particles, the later formed η and β precipitates and other particles. In addition, QOM gives useful quantitative information on the quantities of the particles formed during homogenization. In Fig. 20, as discussed earlier, the increase in the volume fraction of particles during homogenization at low temperatures, is due to the formation of new precipitates (η and β) rather than the increase in the GB particles. On the other hand, the volume fraction of the GB particles is almost unchanged, as shown in Fig. 22. Therefore, the combination of the results from QOM and QXRD quantitative analyses is essential to investigate the evolution of the microstructure during homogenization treatment.

The slight decrease in the fraction of the GB particles during homogenization at low temperatures means that the kinetics of the dissolution of these particles at low temperatures (< 430°C) is relatively slow. Since the fraction of the GB particles decreases significantly during homogenization at 510 °C and higher, it is concluded that in order to dissolve the GB particles, applying a homogenization treatment at 510 °C or higher is necessary.

In the case of QSEM analysis, an unchanged width means that no dissolution has occurred and any decrease in the width of the GB particles is the result of dissolution. The general trend of the change in the width of the GB particles during homogenization using QSEM analysis (Fig. 23) agrees with that of the fractions of the particles from QXRD analysis.

5.3 LMP phases
5.3.1 Results of quantitative analysis using FEG-SEM (QSEM)
Fig. 24 shows the radii of the particles obtained from FEG-SEM after homogenization at various conditions. For these experimental data, EDX analysis was first performed on the

particles which morphologically resembled the LMP phases. The radii of more than 10 particles in each homogenization condition, after ensuring that these were the LMP phases, were measured and the average is reported. The particle radius of zero in this figure indicates that no particles with the composition of the LMP phases were found in the structure, having the same methodology for the prediction of these particles with FEG-SEM at the same magnification. Therefore, the radius of the particles in this case was considered to be zero. It can be observed that the average radius of the LMP phase particles decreases gradually with time during homogenization at 430 and 470 °C.

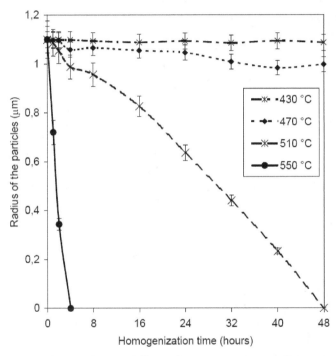

Fig. 24. Average radii of LMP phase at different homogenization conditions in the alloy variant N2 [53]

5.3.2 Results of quantitative differential scanning calorimetry (QDSC)

Fig. 25 shows that after homogenization at a moderate temperature, e.g., 470 °C, the LMP phases are still present in the structure, even after 48 h. At 510 °C, however, the LMP phases are fully dissolved within 48 h and at 550 °C within 2 h, as shown in Fig. 25. This is due to the large increases in the diffusion rates of the elements (i.e., Mg, Cu and Zn) at a high homogenization temperature. Therefore, it is necessary to employ a homogenization treatment at 510 °C for 48 h or 550 °C for 2 h to dissolve the LMP phases.

Fig. 26 gives a close-up view of the peaks in the DSC profiles of the samples homogenized at 470 °C for different hold times. It is clear that the peak intensity decreases with increasing homogenization time up to 48 h, which indicates the decreasing volume fraction of the LMP phases during homogenization at 470 °C.

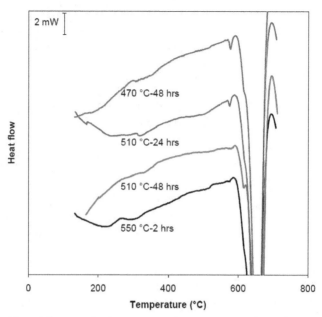

Fig. 25. DCS profiles of the samples homogenized at 470, 510 and 550 °C for different times [53]

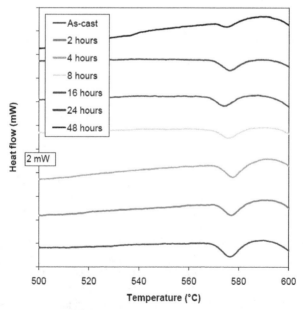

Fig. 26. Regions close to the peaks of the DSC profiles of the samples homogenized at 470 °C for different times (alloy variant N2) [53]

Fig. 27 shows the calculated volume fractions of the LMP phases at different homogenization conditions from the intensities of the DSC peaks, using Eqs. (4) and (5). It is clear that the volume fraction of the dissolved LMP phases increases with time during homogenization. Moreover, the volume fraction of the dissolved LMP phases is larger at a higher temperature and full dissolution only occurs at 510 or 550 °C at reasonable times (less than 48 h).

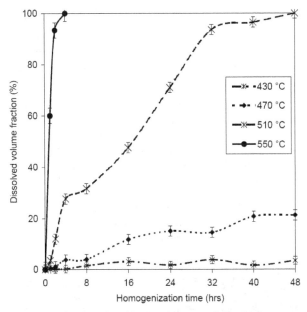

Fig. 27. Volume fractions of the dissolved LMP phases at different homogenization conditions (alloy variant N2) [53]

6. Formation of dispersoids

6.1 Detection of different dispersoid types

Table 8 shows the results of EDX analysis on more than 450 dispersed particles in a size range of less than 100 nm, with different shapes and at different locations with respect to the grains, i.e., grain interior and grain boundary regions in the alloy variant N3 (Table 2) homogenized under various conditions. It was possible to differentiate between 4 types of dispersoids, Zr- (Type 1), Cr- (Type 2), Mn-containing (Type 3) dispersoids and the ones containing a mixture of various elements (Type 4). The number fraction of each dispersoid type, out of 450 dispersoids counted, is also presented in Table 8. It can be seen that 62% of the dispersoids are Zr-containing ones, which indeed account for the majority of the dispersed particles. The number fractions of the other types (2, 3 and 4) are 23, 14 and 1%, respectively. Also shown in Table 8, at the grain boundary regions, i.e., 5 µm from both sides of the grain boundary particles ($Al_{17}(Fe_{3.2},Mn_{0.8})Si_2$ particles), the number fraction of the Mn-containing dispersoids (Type 3) reaches 93 %. However, this type of dispersoids was only observed after homogenization at 510 °C and higher, and especially for a holding time of 4 h or longer. It can be seen that the number fraction of Zr- and Cr-containing dispersoids at the grain boundary regions are only 2 and 1%, respectively.

Dispersoid type	Mg	Fe	Zn	Zr	Cr	Mn	Al	Number fraction (%) out of total 450 dispersoids analyzed	Number fraction (%) out of 75 dispersoids anlyzed at the GB at T≥510°C, time≥4 hrs
Type 1 (Zr-containing)	0.08	0.67	0.1	11.9	0.3	0.09	86.86	62	2
Type 2 (Cr-containing)	0.23	11	0.82	0.6	34.1	0.0	53.22	23	1
Type 3 (Mn-containing)	0.09	7.3	0.26	0.2	0.02	12.8	79.33	14	93
Type 4	4.59	3.2	10.24	0.07	0.12	0.07	81.71	1	4

Table 8. Chemical compositions and number fractions of different types of dispersoids detected in the microstructure of the alloy variant N3

Fig. 28 (a) gives a close-up view of the different types of disperoids. It is clear that the semi-spherical Zr-containing ones (Type 1) are the smallest, while the Cr-containing ones (Type 2) which are fully spherical are the largest. The small sizes of the Zr-containing dispersoids may be attributed to the very low diffusion rate of this element in the aluminum matrix and high nucleation rate [61]. Type 3, also semi-spherical in shape, is not present in the grain interior, thus not shown in Fig. 28 (a). It is presented in Fig. 29. Type 4 which has an elliptic or rod shape morphology is also shown in Fig. 28 (a). To assure the trustfulness of the particles determined with FEG-SEM, TEM analysis was also performed on the same sample and the results are presented in Fig. 28 (b). It can be seen that the same three different types of dispersoids are present in the TEM image with an approximately the same size range.

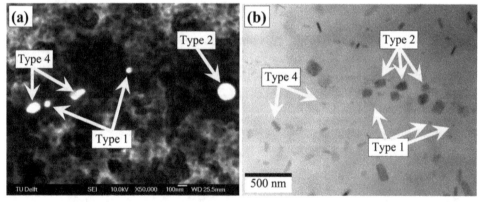

Fig. 28. (a) Close-up view of the different types of dispersoids in a sample homogenized at 510 °C for 8 h and (b) TEM image of the dispersoids in the same sample (alloy variant N3) [62]

Fig. 29 shows the distributions of the different types of dispersoids after homogenization. Fig. 29 (a) and (b) which illustrates the distributions of Zr- and Cr-containing dispersoids (Type 1 and 2) were captured in the grain interior. The number densities of these types decrease significantly with decreasing distance toward the grain boundaries. Fig. 29 (c) illustrates the distribution of Mn-containing dispersoids in the vicinity of an $Al_{17}(Fe_{3.2},Mn_{0.8})Si_2$ particle. It is clear that most of the particles in this region are Mn-containing dispersoids. Their number density inside the grain interior is however almost zero. During the analysis, it was not possible to capture a region with a sufficiently large number of Type 4 dispersoids. Therefore, no figure showing the distribution of this type can be presented in Fig. 29.

Quantitative Zr, Cr and Mn measurements from a line-scan across a grain in the as-cast microstructure of the AA7020 alloy are shown in Fig. 30. It can be seen that the concentrations of Zr and Cr in the grain interior are higher than what would be expected from the peritectic Al-Zr and Al-Cr phase diagrams [65, 66]. The lowest concentrations of Zr and Cr are found at the grain boundaries, which is in agreement with the finding of other researchers [15, 20, 25, 39]. The fluctuations in the Zr concentration across the analyzed region reflect the underlying dendritic structure within each grain. The measurements made close to the centre of the dendrite arms show that the Zr and Cr concentrations exceed their nominal values of 0.2 and 0.1 wt.% in the alloy variant N3, respectively. These regions solidified first during DC-casting and were thus enriched in Zr and Cr. However, a large fraction of grains contained Cr below its nominal value. In particular, low Cr levels were

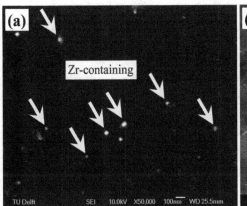
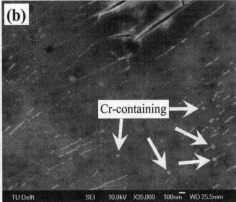
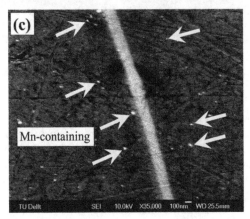

Fig. 29. Distributions of the different types of dispersoids after homogenization at 510 °C for 8 h, (a) Type 1 (Zr-containing), (b) Type 2 (Cr- containing) and (c) Type 3 (Mn-containing) (alloy variant N3) [62-64]

found near the grain boundaries and interdendritic regions. Therefore, it can be concluded that the reason for the small numbers of Cr-containing dispersoids in the grain boundary regions is the negligible concentrations of this element in these regions due to microsegregation during DC-casting. However, the fluctuations in the Zr level are not significant. The lowest Zr level is close to 0.13 wt.%.

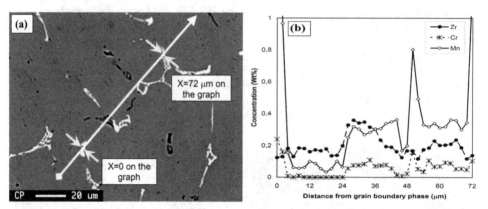

Fig. 30. (a) An EPMA scan and (b) analysis showing segregated Zr, Cr, Mn and Al at the cell boundaries of the as-cast microstructure in the AA7020 aluminum alloy variant N3 [62-64]

As indicated earlier, most of the dispersoids in the grain boundary regions are Mn-containing dispersoids. They are almost absent in the grain interior. The results of the EPMA analysis confirmed that the Mn concentration in the as-cast microstructure in the grain interior is less than the solid solubility limit of Mn at the mentioned homogenization temperatures [48]. Therefore, the reason for no Mn-containing dispersoids formed in the grain interior is a very small concentration of Mn in this region. The peak in the Mn concentration presented in Fig. 30, corresponds to the measurements made on or close to the intermetallic $Al_{17}(Fe_{3.2},Mn_{0.8})Si_2$ particles (GB particles) at the grain boundaries. The small concentrations of Mn in the grain interior and the peaks in Fig. 30 indicate that, during solidification, most of the Mn element localized in the grain boundary regions formed the $Al_{17}(Fe_{3.2},Mn_{0.8})Si_2$ particles located at the grain boundaries.

As mentioned earlier, the Mn-containing dispersoids form only at high temperatures (> 510 °C) and holding times longer than 4 h and in the grain boundary regions. During homogenization at high temperatures, $Al_{17}(Fe_{3.2},Mn_{0.8})Si_2$ particles may dissolve in the microstructure, thus increasing the Mn concentration in the grain boundary regions. Fig. 31 shows the EPMA measurements of the Mn concentration from a line scan across a grain in a sample homogenized at 550 °C for 8 h. It is clear that the Mn concentration in the regions close to the $Al_{17}(Fe_{3.2},Mn_{0.8})Si_2$ particles increases, which is attributed to the dissolution of $Al_{17}(Fe_{3.2},Mn_{0.8})Si_2$ particles during homogenization under this condition. This results in the formation of Mn-containing dispersoids close to the grain boundaries. However, the Mn concentration in the grain interior only changes slightly as confirmed by the present EPMA analysis (Fig. 31), mainly because of the low diffusion rate of Mn in the aluminum matrix [67]. As a result, the Mn concentration is too small in the grain interior to form the Mn-containing dispersoids.

From Fig. 29, it is clear that the Cr-containing dispersoids (Type 2) have a localized distribution in the grain interior. The localized distribution of the Cr-containing dispersoids together with the fact that they are larger than the other types of particles may indicate that this type of dispersoids form heterogeneously, which is in agreement with other investigations [34]. The heterogeneous nucleation of the Cr-containing dispersoids on the u-phases in the case of Al-Mg-Si alloys has been documented [34]. It was observed that during the heating of the as-cast Al-Mg-Si alloys to 580 °C an intermediate phase, referred to as the 'u-phase' nucleated on the Mg_2Si needles. The phase was rich in Mn or Cr. Upon continued heating, the dispersoids containing Mn and Cr nucleated heterogeneously on the 'u-phase' precipitates before these precipitates were dissolved.

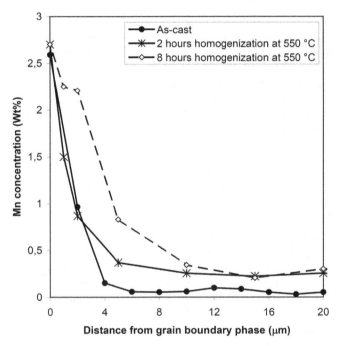

Fig. 31. EPMA scan analysis showing the Mn concentration across a grain of the AA7020 aluminum alloy (alloy variant N3) homogenized at 510 °C for 8 h [62, 64]

To make a solid conclusion on various dispersoids formed during homogenization, they were characterized with respect to formation temperature, size, location and morphology. Table 9 shows the characteristics of the four types of dispersoids formed in the homogenized AA7020 alloy. The deviation expresses the ratio of the number of the particles (with the mentioned characteristics) that showed different chemical compositions to the total number of the particles (with the mentioned characteristic) that were analyzed. The Zr-, Cr- and Mn-containing dispersoids have deviations of only 3, 5 and 4.5 %, respectively. It indicates that the characteristics presented in Table 9 are reasonably accurate to differentiate between these types of dispersoids. It should be noted that the deviation of Type 4 is not reported since the dispersed particles of this type are rare in the microstructure.

Dispersoid type	Major elements	Formation temperature (°C)	Location	Radius (nm)	Morphology	Distribution	Deviation (%) *
Type 1	Zr	390 – 550	Center of the grains	<25	Semi-spherical	Uniform	3
Type 2	Cr, Fe	390 – 550	Center of the grains	25-100	Fully spherical	Localized	5
Type 3	Mn, Fe	T ≥ 510 °C	Close to the GB phases	<50	Semi-spherical	Uniform	4.5
Type 4	Mg, Zn, Fe	390 – 550	All	25-100	Elliptic	Uniform	---

* The deviation expresses the ratio of the number of the particles with the mentioned characteristics that showed different chemical compositions to the total number of analysed particles with the same characteristics.

Table 9. Characteristics of different types of dispersoids in AA7020 after 4 h homogenization at different temperatures

6.2 Evolution of Zr-containing dispersoids during homogenization

As mentioned in Table 8, Zr-containing dispersoids constitute about 62 % of all dispersoids present in the microstructure of the alloy (N3) after homogenization. Therefore, they can be considered the most important ones for recrystallization inhibition. In addition, due to their higher number density, it is easier to quantify them and their evolution during homogenization. Fig. 32 presents typical FEG-SEM images of the Zr-containing dispersoids and related size distribution graphs in the central region of a grain. The size distribution graphs were obtained using the Johnson-Saltykov method as mentioned in the experimental procedure [48] and therefore, the x axis has logarithmic size distribution categories, which is finer at lower values and vice versa.

It should be mentioned that in order to evaluate the efficiency of a homogenization treatment on the inhibition of recrystallization, all relevant parameters including size, size distribution and volume fraction of particles should be taken into consideration, which have been incorporated into an equation of Zener drag pressure [30]. This equation has recently been developed to include the effect of size distribution of dispersoids [3, 68]. The intention of preparing this chapter is to investigate the microstructural evolution during homogenization and therefore, discussion on the effect of homogenization treatment on recrystallization inhibition has been avoided. For more details, the reader is referred to other references [3, 63, 64, 68, 69].

The number density of the Zr-containing dispersoids decreases significantly with increasing distance towards the grain boundaries. This can be attributed to the segregation of zirconium during solidification. It is clear that the sizes and number densities of the Zr-containing dispersoids change with homogenization condition.

Comparison between Figs. 32 (a) and (b) shows that at a given temperature, the sizes increase slightly with time while the number densities remain almost unchanged. In addition, the number of particles of larger sizes, i.e., larger than 15-20 nm, increased with increasing homogenization time at a certain temperature. Comparison between Figs. 32 (b) and (c) for a given composition of the alloy variant N3 and different homogenization temperatures demonstrates that at a higher homogenization temperature e.g., 550 °C, Fig. 32 (c), a larger fraction of particles are of large sizes and the number densities are relatively small, which may be considered as a sign of coarsening. In addition, paying attention to the effect of chemical composition in Fig. 32 (b) (for alloy variant N3) and Fig. 32 (d) (for alloy variant N1), one can find reasonably similar size distributions with different number densities. The sizes are almost constant while the number densities increase significantly with increasing Zr content in the alloy.

Fig. 33 (a) shows the effect of homogenization time on the diameters of the Zr-containing dispersoids at different homogenization temperatures. It is clear that at each homogenization temperature, the average dispersoid diameter increases with increasing holding time and then tends to reach a constant value. In addition, the average diameter is also a function of homogenization temperature. At a higher temperature, the dispersoid diameter is larger.

The effect of homogenization time on the number density of the dispersoids at different homogenization temperatures is illustrated in Fig. 33 (b). It can be seen that with increasing holding time, the number density increases and then stays at a certain level. In addition, the number density of the dispersoids formed at 470 °C is significantly larger than that formed at 390 or 550 °C, while the difference between 390 and 550 °C in the dispersoid number density is considerably smaller.

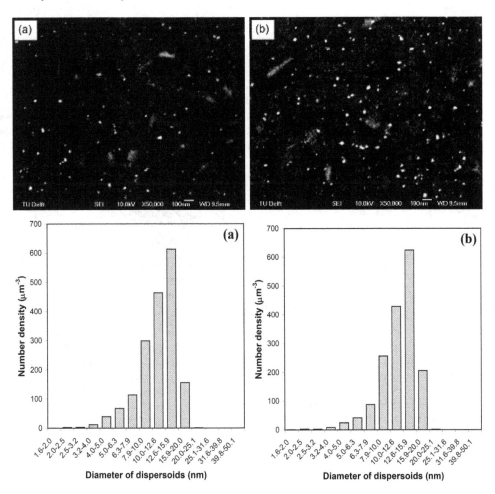

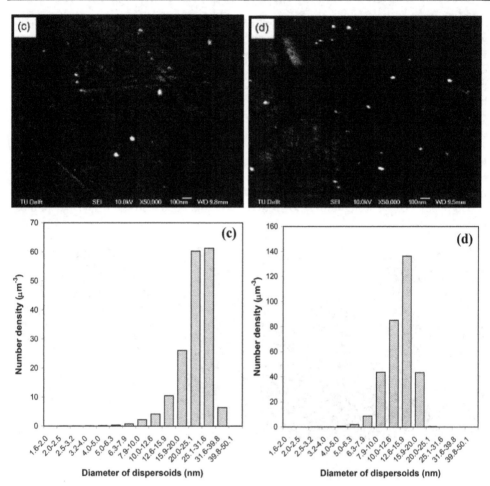

Fig. 32. Typical SEM micrographs showing the effects of homogenization parameters and Zr content on the sizes and size distributions of the Zr-containing dispersoids (a) alloy variant N3, T=470 °C for 8 h, (b) N3, T= 470 °C for 24 h, (c) alloy variant N3, T=550 °C for 24 h and (d) alloy variant N1, T= 470 °C for 24 h [63, 69]

Fig. 34 (a) shows the effect of Zr content on the average diameter of dispersoids formed at 470 °C as a function of time. It can be seen that the average diameter of the dispersoids formed in the alloy with the highest Zr content are larger than those in the other two alloys. However, this effect is not very strong, as the average dispersoid diameters in the N1 and N2 variants do not differ much from each other. An increase in the average diameter of the dispersoid particles with increasing homogenization time towards a constant value is also observed for the alloy variants with different Zr contents. Fig. 34 (b) presents the effect of Zr content on the number density of the Zr-containing dispersoids homogenized at 470 °C. It is clear that the Zr-content has a strong effect on the number density of the dispersoids. The number density of the dispersoids for the alloy with a Zr content of 0.2 wt.% is almost two times as much as that in the alloy with a Zr-content of 0.13 wt.%.

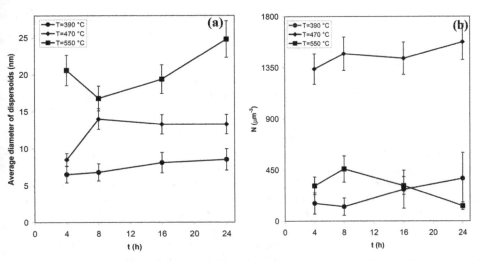

Fig. 33. Effect of homogenization time on the (a) average diameter and (b) number density of the Zr-containing dispersoids in the alloy variant N2 homogenized at different temperatures [62, 69]

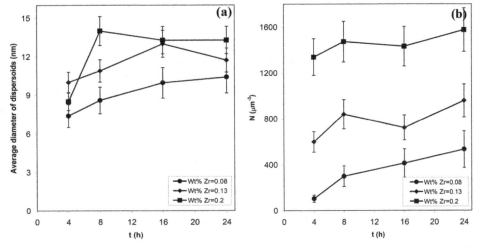

Fig. 34. Effect of Zr content on the (a) average diameter and (b) number density of dispersoids formed at 470 °C as a function of homogenization time [62, 69]

7. Conclusions

Main particles detected in the as-cast microstructure of AA7020 aluminum alloys were categorized to be grain boundary ones, low melting point particles and dispersoids. The evolution of these particles during the homogenization treatment of the AA7020 aluminum alloy was quantitatively analyzed and the following conclusions have been drawn.

1) The particles distributed along the grain boundaries which constitute more than 70% of the secondary phases present in the as-cast structure of the AA7020 aluminum alloy are $Al_{17}(Fe_{3.2},Mn_{0.8})Si_2$ particles. The low melting point phases are indeed present in the as-cast microstructure of the AA7020 aluminum alloy, which may cause incipient melting at 576 °C. These phases contain Al-Cu-Mg-Zn and dissolve during homogenization at 550 °C for 2 h.
2) The width of the grain boundary particles remains unchanged during homogenization at low temperatures. It however decreases at higher temperatures. The extent of the dissolution is more dependent on homogenization temperature than on time. The evolution mechanisms of the GB particles during homogenization consist of spheroidization during homogenization at low temperatures and thinning, discontinuation and full dissolution (TDFD) at high temperatures.
3) Four different types of dispersoids are formed in the AA7020 aluminum alloy variants during homogenization. In addition to the well-known Al_3Zr dispersoids, three other types of dispersoids are also present in the homogenized microstructure of the AA7020 aluminum alloy. The number densities of Zr- and Cr-containing dispersoids are large in the grain interior and very small in the grain boundary regions. These two types of dispersoids appear to be fully spherical and are formed at all the homogenization conditions. The Mn-containing dispersoids form only when the homogenization temperature is equal to or higher than 510 °C and holding time longer than 4 h. The number density of these dispersoids is close to zero in the grain interior but becomes high in the grain boundary regions. The number density and sizes of the Zr-containing dispersoids increase with increasing Zr content of the alloy and homogenization time.

8. Acknowledgments

This research was carried out under the project number MC 4.04203 in the framework of the Research Program of the Materials innovation institute M2i (www.m2i.nl). The authors acknowledge Mr. K. Kwakernaak and Mr. E.R. Peekstok for assistance in microstructure examination.

9. References

[1] ASM Handbook, 10th ed., vol. 14, Metal Forming, ASM International, Metals Park, OH (1992).
[2] T. Sheppard, *Extrusion of Aluminum Alloys*, Kluwer Academic Publishers, Dordrecht (1999).
[3] A.R. Eivani, Modeling of Microstructural Evolution during Homogenization and Simulation of Transient State Recrystallization leading to Peripheral Coarse Grain Structure in Extruded Al-4.5Zn-1Mg Alloy, PhD Thesis, June 2010, Delft, The Netherlands.
[4] N.A. Belov, D.G. Eskin and A.A. Aksenov, *Multicomponent Phase Diagrams: Applications for Commercial Aluminum Alloys*, Elsevier Science, New York (2005).
[5] L.L. Rokhlin, T.V. Dobatkina, N.R. Bochvar and E.V. Lysova, *J. Alloys Compd.* 367, 10 (2004).
[6] S.T. Lim, Y.Y. Lee and I.S. Eun, *Mater. Sci. Forum* 519-521, 549 (2006).
[7] C. Mondal and A.K. Mukhopadhyay, *Mater. Sci. Eng. A* 391, 367 (2005).
[8] R.K. Gupta, N. Nayan and B.R. Ghosh, *Cana. Metall. Q.* 45, 347 (2006).

[9] H. Ahmed, A.R. Eivani, J. Zhou and J. Duszczyk, Proc. Symp. *Aluminum Alloys: Fabrication, Characterization and Application*, TMS Annual Meeting, New Orleans (2008).

[10] N. Parson, and C. Jowett, Proc. 3rd Australasian Pacific Aluminium Extrusion Conference, Sydney, Australlia, (2005).

[11] N. Parson, S. Barker, A. Shalanski and C. Jowett, Proc. 8th Int. Aluminum Extrusion Technology Seminar, vol. 1 (2004).

[12] A.R. Eivani, H. Ahmed, J. Zhou and J. Duszczyk, Proc. Symp. *Aluminum Alloys: Fabrication, Characterization and Application*, TMS Annual Meeting, New Orleans (2008).

[13] J.W. Martin, *Precipitation Hardening*, 2nd ed., Butterworth-Heinemann, Oxford (1998).

[14] A. Jackson and T. Sheppard, Proc. 6th Int. Aluminum Extrusion Technology Conf., Chicago, Aluminum association, Washington DC, vol. 1 (1996).

[15] J.D. Robson and P.B. Prangnell, *Acta Mater.*, 49, 599 (2001).

[16] G.T. Hahn and A.R. Rosenfield, *Metall. Trans. A*, 6, 653 (1975).

[17] D.S. Thompson, *Metall. Trans. A*, 6, 671 (1975).

[18] G.G. Garrett and J.F. Knott, *Metall. Trans. A*, 9, 1187 (1978).

[19] F.Y. Xie, T. Kraft, Y. Zuo, C.H. Moon and Y.A. Chang, *Acta Mater.*, 47, 489 (1999).

[20] B. Morere, C. Maurice and R. Shahani, J. Driver, *Metall. Mater. Trans. A* 32, 625 (2001).

[21] E. Clouet, A. Barbu, L. Lae and G. Martin, *Acta Mater.* 53, 2313 (2005).

[22] Z. Jia, G. Hua, B. Forbord and J.K. Solberg, *Mater. Sci. Eng. A* 444, 284 (2007).

[23] B.L. Ou, J.G. Yang and M.Y. Wei, *Metall. Mater. Trans. A* 38, 1760 (2007).

[24] J.D. Robson, M.J. Jones and P.B. Prangnell, *Acta Mater.* 51, 1453 (2003).

[25] J.D. Robson, *Acta Mater.* 52, 1409 (2004).

[26] J.D. Robson, *Acta Mater.* 52, 4669 (2004).

[27] O.N. Senkov, R.B. Bhat, S.V. Senkova and J. Tatalovich, *Mater. Forum* 28 (2004).

[28] X. Fan, D. Jiang, Q. Meng and L. Zhong, *Mater. Lett.* 60, 1475 (2006).

[29] R. Ciach and B. Dukiet-Zawadzka, *J. Mater. Sci.* 13, 2676 (1978).

[30] F.J. Humphreys and M. Hatherly, *Recrystallization and Related Annealing Phenomena*, 3rd ed., Elsevier Science Inc., Oxford (1995).

[31] M. Peters, J. Eschweiler and K. Welpmann, *Scripta Metall.* 20, 259 (1986).

[32] O. Daaland and E. Nes, *Acta Mater.* 44, 1413 (1996).

[33] Y.J. Li and L. Arnberg, *Acta Mater.* 51, 3415 (2003).

[34] L. Lodgaard and N. Ryum, *Mater. Sci. Eng. A* 283, 144 (2000).

[35] M. Cabibbo, E. Evangelista, C. Scalabroni and E. Bonetti, *Mater. Sci. Forum* 503-504, 841 (2006).

[36] L. Lodgaard and N. Ryum., *Mater. Sci. Tech.* 16, 599 (2000).

[37] D.H. Lee, J.H. Park and S.W. Nam, *Mater. Sci. Tech.* 15, 450 (1999).

[38] R.A. Jeniski, B. Thanaboonsombut and T.H. Sanders, *Metal. Mater. Trans. A* 27, 19 (1996).

[39] J.D. Robson, *Mater. Sci. Eng. A* 338, 219 (2002).

[40] S.V. Senkova, O.N. Senkov and D.B. Miracle, *Metal. Mater. Trans. A* 37, 3569 (2006).

[41] O.N. Senkov, R.B. Bhat, S.V. Senkova and J.D. Schloz, *Metal. Mater. Trans. A* 36, 2115 (2005).

[42] A. Deschamps and Y. Bréchet, *Mater. Sci. Eng. A* 251, 200 (1998).

[43] J.T. Armstrong, in K.F.J. Heinrich, D.E. Newbury (Eds), *Electron Probe Quantitation*, Plenum Press, New York (1991).

[44] B.D. Cullity, S.R. Stock and S. Stock, *Elements of X-Ray Diffraction*, 3rd ed., Prentice Hall, New Jersi (2001).

[45] Alphabetical indexes for experimental patterns, Sets 1-52, International Center for Diffraction Data (ICDD), 2005, Powder Diffraction File Number (PDF no.) 01-071-4015.

[46] M. Cooper, *Acta Crystallogr.* 23, 1106 (1967).

[47] ASM Handbook, 10th ed., vol. 2, *Non-Ferrous Alloys*, ASM International, Metals Park, OH (1992).

[48] R.T. DeHoff and F.N. Rhines, *Quantitative Microscopy*, McGraw-Hill, New York (1968).

[49] A.R. Eivani, H. Ahmed, J. Zhou, J. Duszczyk, Metal. Mater. Trans. A. 40, 717 (2009).

[50] R.W. Cahn and P. Haasen, *Physical Metallurgy*, North Holland, Amsterdam (1996).

[51] D. Gaskell, *Introduction to the Thermodynamics of Materials*, Taylor & Francis Co., New York (2003).

[52] R. Nadella, D.G. Eskin, Q. Du and L. Katgerman, *Prog. Mater. Sci.* 53, 421 (2008).

[53] A.R. Eivani, H. Ahmed, J. Zhou, J. Duszczyk, Mater. Sci. Tech. 26, 215 (2010).

[54] M. Dumont, W. Lefebvre, B. Doisneau-Cottignies and A. Deschamps, *Acta Mater.* 53, 2881 (2005).

[55] G. Sha and A. Cerezo, *Acta Mater.* 52, 4503 (2004).

[56] A. Melander and P.A. Persson, *Acta Metall.* 26, 267 (1978).

[57] ASM Handbook, 10th ed., vol. 3, *Alloy Phase Diagrams*, ASM International, Metals Park, OH (1992).

[58] M. Conserva, E. Di Russo and A. Giarda, *J. Metall.* 6, 367 (1973).

[59] E. Ho and G.C. Weatherly, *Acta Metall.* 23, 1451 (1975).

[60] A.R. Eivani, H. Ahmed, J. Zhou, J. Duszczyk, Phil. Mag. 42, 1109 (2010).

[61] M. S. Zedalis and M. E. Fine, *Metall. Trans. A.* 17, 2187 (1986).

[62] A.R. Eivani, H. Ahmed, J. Zhou, J. Duszczyk, C. Kwakernaak, Mater. Sci. Tech. DOI 10.1179/026708310X12635619988267 (2010).

[63] A.R. Eivani, H. Ahmed, J. Zhou, J. Duszczyk, Mater. Sci. Eng. A. 527, 2418 (2010).

[64] A.R. Eivani, S. Valipour, H. Ahmed, J. Zhou, J. Duszczyk, Metal. Mater. Trans. A. 40, 2435 (2009).

[65] J. Murray, A. Peruzzi and J.P. Abriata, *J. Phase Equilib.* 13, 227 (1992).

[66] N. Saunders, *Z. Metallkd* 80, 894 (1989).

[67] Y. Du, Y.A. Chang, B. Huang, W. Gong, J. Jin, H. Xu, Z. Yuan, Y. Liu, Y. He and F.Y. Xie, *Mater. Sci. Eng. A* 363, 140 (2003).

[68] A.R. Eivani, S. Valipour, H. Ahmed, J. Zhou, J. Duszczyk, Metal. Mater. Trans. A. 42, 1109 (2011).

[69] A.R. Eivani, H. Ahmed, J. Zhou, J. Duszczyk, Adv. Mater. Res. 89-91, 177 (2010).

7

Nanostructure, Texture Evolution and Mechanical Properties of Aluminum Alloys Processed by Severe Plastic Deformation

Abbas Akbarzadeh
Department of Materials Science and Engineering, Sharif University of Technology,
Tehran,
Iran

1. Introduction

Various research works have been conducted to replace heavy steel body constructions with lighter aluminum ones to achieve stronger energy consumption and environmental standards. The most important technical obstacle to this goal is the inferior ductility of most aluminum sheet alloys. It has been reported that control of the microstructure and the texture of materials is essential for improvement of their mechanical properties (Lee et al., 2002). Reducing the grain size of polycrystalline metallic materials to the nanosize ($d < 100$ nm, nanocrystalline) or submicron levels (100 nm$< d <1$ µm, ultra-fine grain) is an effective and relatively economic way of improving mechanical properties such as strength, toughness, or wear resistance in structural materials (Kim et al., 2006; Prangnell et al., 2001) which even can give rise to superplastic behavior under appropriate loading conditions (Pérez-Prado et al., 2004). Since it is practically difficult to reduce the grain size of many metallic materials such as aluminum alloys below 5 µm by a conventional cold working and recrystallization process, several new methods are developed to manufacture ultrafine grained (UFG) materials (Kim et al., 2006). These methods can be classified into two main groups namely bottom-up and top-down processes. In the bottom-up procedures, such as rapid solidification, vapor deposition and mechanical alloying, an ultra-fine microstructures is configured from the smallest possible constituents which are prohibited to grow into the micrometer domain (Pérez-Prado et al., 2004). In the top-down procedures, on the other hand, an existing microscale microstructure is refined to the submicrometer scale, e.g. by a process such as severe plastic deformation (SPD) (Pérez-Prado et al., 2004; Saito et al., 1999). The ancient Persian swords are the interesting examples of severe upset forging for development of fine microstructures (Sherby and Wadsworth, 2001).

By now, various SPD processes such as accumulative roll bonding (ARB) (Saito et al., 1999), cyclic extrusion compression (CEC) (Richert J. & Richert M., 1986), equal channel angular pressing (ECAP) (Valiev et al., 1991), and high pressure torsion (HPT) (Horita et al., 1996) have been proposed and successfully applied to various materials. The common feature of these techniques is that the net shape of the sample during processing is approximately constant, so that there is no geometric limitation on the applied strain (Prangnell et al., 2001). Among these processes, accumulative roll bonding has some unique features. Firstly, unlike the ECAP, CEC and HPT processes which require forming machines with large capacity and expensive dies, the ARB process can be performed by a conventional rolling mill without any special die. Secondly, in comparison to the other methods, the productivity of the ARB process is relatively high because this process implies the potential of industrial up scaling to a continuous production of UFGed metallic sheets or plates (Saito et al., 1999). In principle at least, the use of heavy deformation in metal processing, with the objective of producing metal alloys with superior properties, has a long history which may be traced back to the early metal-working of ancient China [16], the blacksmith's production of high-quality Damascus steel in the Middle East [17] and the fabrication of the legendary Wootz steel in ancient India [18].

The ARB process is a repetitive procedure of cutting, stacking and roll bonding of similar sheets for a desired number of cycles, Fig. 1, which is practically limited by technological constraints such as the occurrence of edge cracks. In this process the thickness of the sheet varies between fixed limits and by repeating the procedure very high strains can be accumulated in material, and as a result significant structural refinement can be achieved (Huang et al., 2003; Saito et al., 1999; Tsuji et al., 2003a).

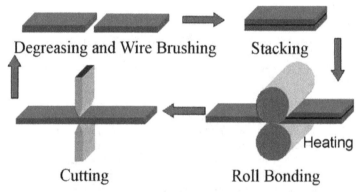

Fig. 1. Schematic illustration showing the procedure of the ARB process (Pirgazi et al., 2008a)

The first scientific paper on the ARB was published in 1998 and afterwards extensive studies have been conducted regarding the microstructural evolution and mechanical properties of various materials processed in different ARB cycles. Pure aluminum (Huang et al., 2003; Saito et al., 1998; Tsuji et al., 2003a), AA5083 (Saito et al., 1999), AA6061 (Lee et al., 2002; Park et al., 2001), AA8011 (Kim et al., 2002, 2005), AA3103 (Chowdhury et al., 2006a) and AA8090 (Chowdhury et al., 2006b) aluminum alloys, magnesium alloys (del Valle et al., 2005; Pérez-Prado et al., 2004), and Ti-IF steel sheets (Reis & Kestens, 2005, Reis et al., 2005; Tsuji et al., 2002a) are the most important materials which have already been successfully produced by the ARB process. The results of these investigations mainly indicate that during the first stages of ARB, ultra-fine grains with diameter less than 1 μm partially form in the sheets, and the volume fraction of these grains increases with increasing the number of cycles, so that after high levels of strain the sample is completely covered with ultra-fine lamellar grains which are not equi-axed and represent an aspect ratio bigger than one (Huang et al., 2003; Park et al., 2001; Saito et al., 1998; Tsuji et al., 2003a). The investigations on the mechanical properties by previous researchers show that the ARB is a promising process for improving this feature of metallic sheets. It has been reported that a significant increase in strength and hardness, more than two or three times of the values of starting materials can be achieved by ARB process in aluminum alloys and IF steel sheets (Lee et al., 2004; Park et al., 2001; Saito et al., 1998; Tsuji et al, 2002b).

Microstructural evolution during ARB has been studied by several researchers. TEM investigation and local crystallography were used (Huang et al., 2003) to study the microstructural evolution of AA1100 up to eight cycles of ARB. Their results show that at large strains, almost a homogenous lamellar structure is formed across the thickness. This microstructure is subdivided by high angle boundaries and low angle dislocation boundaries. The spacing of dislocation boundaries decrease and their misorientations increase by accumulating strain. EBSD analysis illustrates three mechanisms of UFGs development at different levels of strains (Pirgazi et al., 2008b). It was shown that development of subgrains is the major mechanism during the first two cycles of ARB. This mechanism is followed by strain induced transition of low angle grain boundaries to high angles and formation of a thin lamellar structure at medium levels of strain. Fragmentation of these thin lamellar structures into more equi-axed grains is considered as the dominant mechanism after sixth cycle of ARB. A similar microstructural evolution during ARB of AA6061 has been reported (Park et al., 2001). In addition to fabrication of UFGed and nanostructured materials for metallic sheets, some other unique features for ARB process are applied to different sheets to fabricate multi-layer composites (Min et al., 2006).

The texture analysis is a powerful tool to investigate the microstructural and substructural evolution of plastically deformed materials because it provides information on the fragmentation behavior of grains. When orientation contrast microscopy is used, the microstructural data can be linked with local textural information. It is commonly known that in materials produced by severe plastic deformation processes, the band contrast of the corresponding Kikuchi-lines decreases as a result of the presence of lattice defects, such as regions with high dislocation density, subgrains and grain boundaries. In the TSL® software, the contrast of the Kikuchi lines is quantified by the image quality (IQ) parameter. The IQ parameter is sensitive to a wide variety of additional material and instrumental factors which makes it almost impossible to deconvolute the IQ signal in order to convert it to univocal quantitative information on the local microstrain. Nevertheless, the EBSD technique is the only possible tool for nanoscale analysis of relatively large areas in severely deformed materials at present (Tsuji et al., 2002a).

It has been reported (Chowdhury et al., 2006a) that the texture development in an AA3103 alloy during accumulative roll bonding process shows symmetry at all stages and the major components can be characterized as the Dillamore {4411} <11 11 8> component along with the S component with a scatter around the brass component. It has been also reported that in AA8011 aluminum alloy sheets processed by ARB, the deformation texture is dominated by the Dillamore component and the shear texture was developed near the surface of the sheets. This surface shear texture disappears rapidly as the surface area of the material re-appears in the center of the composite sample in the next ARB pass (Kim et al, 2002, 2005). Similar results have been published by (Heason & Prangnell, 2002a). They have reported that in AA1100 alloy processed by the ARB, coarse unrefined bands can be retained even at very high strains. They have also proposed that the strong texture developed during ARB processing may lead to this inhomogeneity (Heason & Prangnell, 2002a). It has also been reported that the textures in Ti-IF steel produced by ARB display a conventional cross sectional gradient with typical shear component in the subsurface planes and plane strain compression components in the midsection (Reis & Kestens, 2005; Kolahi et al., 2009).

It has been reported that in AA8011 aluminium alloy sheets processed by ARB, the main deformation texture orientation is the Dillamore {4 4 11}<11 11 8> component and the shear texture is developed on the surface of the sheets (Kim et al., 2002). This surface shear texture disappears rapidly as the surface area of the material reappears in the centre of the composite sample during the next ARB cycle. The role of second phase particles on the grain size reduction of aluminium sheets during the accumulative roll bonding process has been investigated by comparing the microstructure and texture of a single phase (AA1100) and a particle containing aluminium alloy (AA3003) during various ARB cycles (Pirgazi & Akbarzadeh, 2009).

Circular shaped hollow sections like tubes and cylinders, as a category of engineering components, are also expected to achieve high strengths by nanostructure. Tube spinning is used as a common tube fabricating process (Wong et al., 2003), based on which a cold-bonding process titled "spin-bonding" with the advantages of ARB of sheets has been recently proposed (Mohebbi & Akbarzadeh, 2010a, 2010b) for manufacturing of high strength thin-walled tubes and cylinders. The SPD process proposed in that work is, in fact, repeatedly spin-bonding of layered tubes (accumulative spin-bonding, ASB) to induce large plastic strain on tubes similar to the ARB of sheets, Fig. 2 (Mohebbi & Akbarzadeh, 2010a).

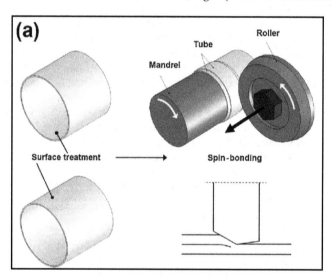

Fig. 2. Schematic illustration of the spin-bonding (Mohebbi & Akbarzadeh, 2010a)

In many cases, the inner and outer surfaces of hollow sections are exposed to different environments, and different characteristics are required inside and outside. In these cases, various bimetallic or clad tubes of stainless steels or high-alloy steels and super alloys clad onto carbon or low alloy steels, are utilized in boilers, heat exchangers, nuclear power plants and petroleum and chemical industries (Chen et al., 2003). So far, several methods have been used to produce composite tubes and cylinders. While centrifugal casting (Sponseller et al., 1998) and extrusion (Chen et al., 2003) can be used to fabricate the thick-walled bimetallic tubes, explosive bonding (Berski et al., 2006), ball attrition (Zhan et al., 2006), thermo-hydraulic fit method and hydraulic expansion method (Wang et al., 2005) are capable of manufacturing thin-walled cylinders. The most widely used cladding process is roll-bonding of two or more sandwich sheets (Zhan et al., 2006). Although it is possible to fabricate seamed composite tubes by sandwich sheets, it is ideal to plan a method with advantages of roll-bonding to produce seamless thin-walled tubes and cylinders. The spin-bonding method in cold-bonding of cylinders was also proposed by utilizing the tube spinning process for manufacturing of clad tubes (Mohebbi, 2009a).

Tube spinning is an incremental and localized deformation in which material between the roller and the mandrel gradually deforms to the final thickness. Actually the deformation area is limited to a part of work piece which is in contact to the roller, so that the deformation is constrained strongly by surrounding metals. This is why the stress and strain have such a complicated distribution in this incremental process. There is always an inhomogeneous material flow due to the local deformation between the roller and the preform surface with a high strain rate (Mohebbi & Akbarzadeh, 2010b, 2010c, 2009b).

2. Experimental methods

The materials used to study the ARB process were fully annealed sheets of Al-Mn alloy (AA3003) and commercial purity aluminum (AA1100) with initial grain sizes of 40 and 34 μm. Thickness of the sheets was respectively 0.4 and 0.5 mm, and the chemical compositions are presented in Table 1. Two sheets of 150×50 mm2 were degreased (in acetone) and wire brushed (by a stainless steel brush with wires of 0.4 mm in diameter). After the surface treatment, the two sheets were stacked on top of each other and preheated to a temperature of 250 °C for 5 minutes. The plane strain rolling was performed along the longest dimension by 50% reduction in thickness at 200 °C without any lubrication and the mean strain rate was 51 sec^{-1}. Afterwards the roll bonded sample was cut into two sheets of approximately the initial dimension and the procedure was repeated up to a total of 8 cycles so that an accumulated equivalent strain of 6.4 was achieved.

Alloy	Fe	Cu	Si	Ti	Mg	Mn	Al
AA1100	0.57	0.12	0.13	0.03	0.02	0.013	Bal.
AA3003	0.436	0.216	0.18	0.009	0.005	1.076	Bal.

Table 1. Chemical composition of the aluminium alloys (Pirgazi & Akbarzadeh, 2009)

The microtexture and microstructure measurements were performed on the section which is perpendicular to the transverse direction of rolling and is located at the mid-thickness of the sheets. The analysis was carried out by employing an Orientation Image Microscopy (OIM) attached on a Philips XL30 ESEM microscope equipped with a LaB6 filament. The electron back-scattering diffraction (EBSD) mappings were carried out with step sizes in the range of 0.12 μm to 80 nm and the OIM software developed by TSL® was used for data acquisition and post-processing procedures.

Since the ARBed samples were severely deformed, acquiring good Kikuchi-patterns was not evident. By means of electro polishing a very smooth surface was produced and by adjusting the operating parameters of the SEM and the software (OIM Data Collection) a satisfying result could be obtained. In this way, after using a grain confidence index (CI) standardization (5°, two points) and neighbor orientation correlation (CI = 0.1) cleanup procedures on the raw data, confidence indices of 0.9 and 0.79 were obtained for the samples processed by 4 and 10 cycles, respectively.

The Vickers micro hardness test was utilized to investigate the mechanical properties of initial and ARB processed samples. The values reported for HV represent the average of ten measurements taken at randomly selected points across the thickness of the sheets using loads of 50 and 100 g for 20 s. The mechanical properties of initial and ARBed samples were measured by tensile tests at room temperature executed with an Instron tensile testing machine. The test specimens were prepared with the tensile axis parallel to the rolling direction. The test conditions and the specimen size were chosen according to the ASTM-E8 standard.

Orientation distribution functions (ODFs) were calculated by harmonic series expansion with truncation at L = 16. The initial texture of the fully annealed AA3003 alloy sheet prior to the ARB process was also characterized using the EBSD analysis. In order to obtain a statistically acceptable result, this measurement was carried out with a step size of 2 μm over an area with dimension of 300 μm×300 μm located at mid thickness and as a result more than 22,000 orientations were analyzed.

The texture evolution during ARB process was predicted with the Alamel model (Van Houtte et al., 2005, 2006). In this model the texture is discretized in a set of N individual orientations which are considered in pairs of two. Each pair of orientations will accommodate the externally imposed strain but shear strains can be relaxed in equal and opposite measure for each grain belonging to a pair. In the ODF calculations the orthorhombic sample symmetry was imposed which is usually assumed for the conventional rolling process. Only the {111} <110> slip systems were taken into account for the texture calculations.

The tubes used for ASB process were prepared from commercially pure aluminum (AA1050). Thickness of the tubes was 0.8 mm. Work pieces were annealed at 350 °C for 2 hrs before the first cycle. Spin-bonding process was repeated up to four cycles. In the spin bonding process two surfaces to be bonded (inner surface of the external tube and outer surface of the internal one) were degreased in acetone and wire brushed as surface treatment. After scratch brushing, surface treated tubes were positioned against each other and fitted on the mandrel for tube spinning at room temperature with conditions of Table 2 (Mohebbi & Akbarzadeh, 2010c). At this stage, while the tube and mandrel rotate about their axes, a roller with a degree of freedom about its own axis moves along the direction of the tube axis to reduce its thickness to 50% leading to bonding of the tubes. No preheating was performed in this work. More details of the spin-bonding is available in (Mohebbi & Akbarzadeh, 2010a).

The longitudinal sections of the specimens were observed by an optical microscope after polishing and etching in the Tucker solution for 15 s. Transmission electron microscopy (TEM) micrographs and the corresponding selected area diffraction (SAD) patterns were also obtained from the specimens after 1, 2 and 4 cycles of ASB. To do so, thin foils were prepared by twin-jet polishing from the mid-thickness of the tubes normal to the tube surface. The microstructural evolution was also analyzed by electron backscattering diffraction (EBSD) pattern. This was performed on the longitudinal sections perpendicular to the peripheral direction of the tubes after various cycles of ASB. The measured area was 25 μm×30 μm located at the center of mid-thickness of the tubes.

Parameter	Value
Workpiece	
Inner diameter (mm)	35
Wall thickness (mm)	2.5
Initial length (mm)	50
Roller	
Diameter (mm)	54
Attack angle (°)	25
Smoothing angle (°)	5
Flow forming conditions	
Feed rate (mm/rev)	0.1
Reduction (%)	40
Speed of rotation (rpm)	30

Table 2. Dimensions of tools and work piece and the process conditions

EBSD patterns were obtained at 25 kV and analyzed by TSL software regarding the quantitative analysis of grain boundaries and misorientation distributions. The step size was 100 nm for the specimens after the first and second cycles and 60 nm for the specimens after the third and fourth cycles of ASB. Mechanical properties of the specimens were determined at ambient temperature with strain rate of 2.6×10^{-3} sec^{-1}. The samples were prepared from the circumferential direction of the tubes after straightening and machining according to the ASTM E8M standard. Vickers microhardness values were also measured through the thickness of the specimens by applying the load of 0.49 N.

The material used in ASB process was the commercial purity aluminum (AA 1050). Thickness of the tubes was 0.8 mm. The inner diameter of the internal tube and the external one were 51 and 52.8 mm, respectively and their length was 40 mm. Work pieces were annealed at 350 °C for 2 hrs before the bonding process. Since they were processed via tube spinning, work pieces were diametrically true after annealing. The bonding surfaces were degreased in acetone and scratch brushed. The time between surface preparation and bonding process was kept to less than 300 s to minimize the formation of contaminant film and a thick oxide layer on the bond surfaces of the tubes. Afterwards, the two tubes to be joined were positioned against each other. The process is schematically illustrated in Fig. 2.

The bond strength of the Al/Al bimetal layers was measured using the T-peel test. Samples were 8 mm wide and 50 mm long and were cut at the longitude of the tubes so that their width was in the circumferential direction. The breaking off force per width of the sample was measured as the bond strength (N/mm). Optical microscopy was employed to examine the bond interface of the longitudinal section of deformation zone. The surfaces to be examined were prepared by standard metallographic procedure to polished conditions without etching.

3. Results and discussions

3.1 Microstructure
3.1.1 ARB processing
Figure 3 shows the optical micrograph observed in the RD-ND section of the sample produced by 10 ARB cycles, indicating that a good bonding with no delamination has been achieved under the present ARB conditions. To investigate the microstructural evolution of

aluminium sheets processed by ARB, the boundary misorientation maps were extracted from the EBSD data. These mappings were recorded on a section near the centre of the RD–ND plane of the samples. Because of the severe deformation of the ARB processed samples, acquiring good Kikuchi-lines was too difficult. However, by means of electropolishing in Barker's reagent (5 mL HBF4 in 200 mL H2O) and preparing a very smooth surface and adjusting the parameters of SEM (e.g. gun high voltage, working distance and spot size) and the OIM Data Collection software (e.g. exposure time, binning and step size), it was significantly improved.

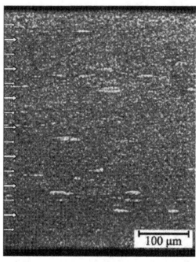

Fig. 3. Optical micrograph of longitudinal cross section of aluminum sheet (AA1100) after 10 ARB cycles

The orientation scans of the AA1100 and AA3003 aluminum sheets processed by various ARB cycles are depicted in Figs. 4 and 5. The boundary misorientation maps of these samples are also depicted in these figures. In these maps, the high angle grain boundaries (HAGB) with misorientations larger than 15° are drawn in bold black lines, while the low angle grain boundaries (LAGB) with misorientations between 2° and 15° are drawn in thin grey lines. In the case of AA1100 alloy, it is observed that after two cycles of ARB, the microstructure is covered with elongated grains surrounded by many low angle grain boundaries (Fig. 4a). The microstructures of the samples produced by four and six ARB cycles are generally similar to the sample produced by two ARB cycles. However, with the increasing number of cycles the grain sizes decrease gradually. After the sixth cycle, the structure consists of very fine elongated grains surrounded by high angle grain boundaries which are usually parallel to the rolling direction (Fig. 4c).

It is observed that the evolution of ultra-fine grains in the AA1100 alloy occurs by various mechanisms of grain refinement at different strains. At low strains (ε<1.6) grain subdivision is the dominant mechanism which leads to the formation of a fibrous microstructure of elongated grains and promotes the formation of subgrains surrounded by a high fraction of low angle grain boundaries (LAGB) (Fig. 4a) and some coarse unrefined bands which contain predominantly subgrains are seen in this specimen. At medium levels of strains

(1.6<ε<4.8) the grain subdivision ratio decreases and strain induced transformation of low angle to high angle grain boundaries (HAGB) caused by the accumulation of dislocations in exiting subgrain boundaries is the controlling mechanism which results in a lamellar structure at the specimens processed by 4 and 6 ARB cycles (Figs. 4b and c). The fraction of HAGB presented in Figs. 4 and 5 confirm the increase of the HAGBs in expense of the LAGBs.

In comparison with the first sample, less pronounced unrefined regions of subgrains are also seen in these samples. The proposed mechanism of grain refinement at high levels of strain (ε>4.8) is the progressive fragmentation of the thin lamellar grains into more equi-axed grain structures which leads to development of a relatively homogeneous submicron grain structure without the unrefined regions after the eighth cycle (Fig. 4d).

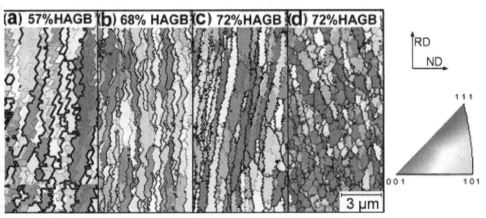

Fig. 4. Orientation scans obtained from the EBSD measurements of the AA1100 ARB processed by (a) 2, (b) 4, (c) 6 and (d) 8 cycles. Thin grey lines represent the misorientation (θ) of 2° ≤θ <15° and bold black lines represent 15° ≤ θ

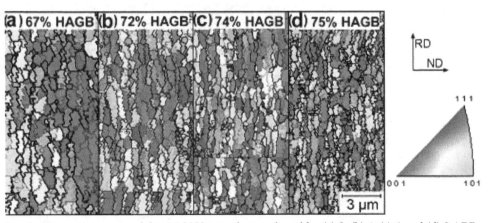

Fig. 5. Orientation scans of the AA3003 samples produced by (a) 2, (b) 4, (c) 6 and (d) 8 ARB cycles

Fig. 5 displays that in comparison with the AA1100 alloy, in the AA3003 alloy which contains a significant volume fraction of second phase particles, an ultra-fine grain structure develops at a much higher rate. It is reported that two types of precipitates evolve in the AA3003 alloy: orthorhombic (Fe,Mn)Al6 and cubic α-Al(Fe,Mn)Si. The total volume fraction of these precipitates is between 4 and 5% and the large majority of them are elongated and have an average length between 1-5 μm (Richert J. & Richert M., 1986). In a similar research on AA8097, it has been shown that these second phase particles cause the texture randomization and promotion of grain refinement (Heason and Prangnell, 2002b).

According to Fig. 5, the volume fraction of HAGBs after the second cycle in the AA3003 alloy is 67% compared to 57% in the AA1100 alloy. This rapid refinement of grains in the AA3003 alloy at relatively low strains results from extensive HAGBs generation in local deformation zones around the second-phase particles. During deformation, large lattice rotation around the second phase particles leads to the increase of local misorientation and production of new HAGBs. Dispersion of these particles also develops a random and more heterogeneous plastic flow behavior within the matrix and promotes the fragmentation of the lamellar structures seen in the AA1100 alloy (Pirgazi & Akbarzadeh, 2008c). This behavior as well as a gradual increase in the percentage of HAGBs and a reduction in the grain aspect ratio during the next cycles results in the formation of a more homogenous submicron grains structure in the AA3003 alloy (Fig. 5).

Although many low angle grain boundaries are also observed in the microstructure, they are usually perpendicular to the rolling direction. Figure 4d displays the microstructure developed in the AA1100 aluminium sheet processed by eight ARB cycles. Significant changes are observed at this level of strain and the grain morphology changes into more equi-axed structures. In comparison with the AA1100 alloy, the developed microstructures in AA3003 alloy during the ARB process are somewhat different. According to Fig. 5, it is clearly observed that in this alloy, the formation of ultrafine grains occurs at the earlier stages of the process. Figure 5a shows that even after the second cycle, the microstructure is completely covered with very fine grains whose morphology is more equi-axed than the same sample of AA1100 alloy (Fig. 4a).

With the number increasing of cycles, the grain size continuously decreases and after the eighth cycle the whole specimen shows a homogenous submicrometre grain size structure. Further analysis revealed that the microstructures of these samples differ in grain size and the fraction of HAGBs. Variations of microstructural parameters of the aluminium sheets during the ARB process are depicted in Fig. 6 with regard to the changes of the average thickness and length of grains. In the AA1100 alloy, the grain thickness (measured by the linear interception method on the EBSD maps) drastically decreases during the first four cycles. While with the increasing strain during the next cycles, the grain thickness reduces slightly and approached to a constant value of about 500 nm at large strains. Similar results are observed in the AA3003 alloy but the grain thickness in this case is smaller than the former case and reaches the minimum of 400 nm after the eighth cycle. Comparison of the grain length in the ARB processed samples revealed that there is a more dramatic difference between the evolved microstructures in the two alloys after a large number of ARB cycles. In the AA3003 alloy, the average length of grains decreases continuously while in the case of AA1100 alloy a drastic reduction is observed after the sixth cycle (Fig. 6b).

By analysing the EBSD data after each ARB cycle, quantitative measurements were made of the misorientation distributions and the results are summarised in Fig. 7. The Mackenzie distribution for boundaries in a random polycrystal is shown by the black bold line. In both alloys, with ARB processing, the fraction of high angle grain boundaries increases and approaches to a near random misorientation distribution very much similar to the theoretical Mackenzie distribution. However, there are some differences between the misorientation profiles produced in the two alloys during the ARB process. It can be observed that in AA3003 alloy the fraction of HAGBs initially increases rapidly and reaches over 70% after the fourth cycle. During the next cycles, few further changes occur and the fraction of HAGBs after the eighth cycles reaches to 75%. In contrast, the fraction of HAGBs in AA1100 alloy increases more slowly with the increasing strain during the ARB process and reaches a saturation value of about 72% after the sixth cycle.

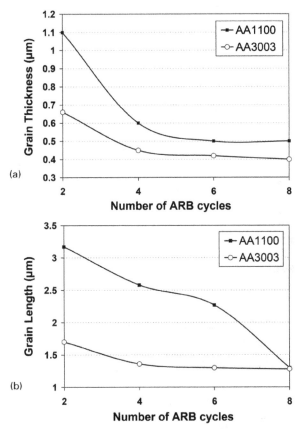

Fig. 6. Variation of microstructural parameters of aluminium sheets during ARB process, (a) grain thickness and (b) grain length

A general definition for the materials with a submicrometer grain structure proposed by (Prangnell et al., 2001) includes two main criteria: the average spacing of HAGBs must be less than 1 μm, and the fraction of HAGBs must be greater than 70%. The authors' observations proved that the development of such submicrometer or ultrafine grains in a single phase (AA1100) and a particle containing alloy (AA3003) occurs at different levels of strains during the ARB process. It was shown that (Heason & Prangnell, 2002a) the evolution of ultrafine grains in the AA1100 alloy occurs by various mechanisms of grain refinement at different strains.

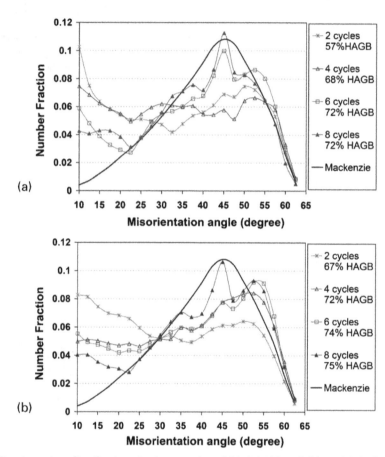

Fig. 7. Misorientation distributions in the samples of (a) AA100 and (b) AA3003 aluminium sheets processed by various ARB cycles

At low strains (ε< 1.6), grain subdivision is the dominant mechanism which leads to the formation of a fibrous microstructure of elongated grains. This mechanism also promotes the formation of subgrains surrounded by a high fraction of low angle grain boundaries (Fig. 4a) and some coarse unrefined bands which contain predominantly subgrains are seen in this specimen. At medium levels of strains ($1.6 < \varepsilon < 4.8$) the grain subdivision ratio decreases and the strain induced transformation of low angle to high angle grain boundaries caused by the accumulation of dislocations in exiting subgrain boundaries is the governing mechanism which results in a lamellar structure at the specimens processed by four and six ARB cycles (Figs. 4b and c).

The misorientation profiles plotted in Fig. 7 confirm the increase of HAGBs in expense of the LAGBs and the tendency of misorientation of the lamellar structure to a Mackenzie distribution. In comparison with the first sample, less pronounced unrefined regions of subgrains are also seen in these samples. The proposed mechanism for grain refinement at high levels of strain ($\varepsilon > 4.8$) is the progressive fragmentation of thin lamellar grains into more equi-axed grain structures (Fig. 4d). The last mentioned mechanism enables the ARB process to develop a relatively homogeneous submicrometer grain structure in aluminium sheets.

Figure 5 displays that in comparison with the AA1100 alloy, in the AA3003 alloy which contains a significant volume fraction of second phase particles, an ultrafine grain structure develops at a much higher rate. It has been reported that two types of precipitates evolve in the AA3003 alloy: orthorhombic $(Fe,Mn)Al_6$ and cubic α-$Al(Fe,Mn)Si$ (Rios & Padilha, 2003). The total volume fraction of these precipitates is between 4 and 5% and the large majority of them are elongated and have an average length between 1 and 5 mm. According to Fig. 7, the volume fraction of HAGBs after the second cycle in the AA3003 alloy is 67% compared with 57% in the AA1100 alloy. At the same time, the transverse spacing of HAGBs in these specimens is 0.66 and 1.1 mm respectively (Fig. 6a). This rapid refinement of grains in the AA3003 alloy at relatively low strains results from extensive generation of HAGBs in local deformation zones around the second phase particles. During deformation, large lattice rotation around the second phase particles leads to the increase of the local misorientation and production of new HAGBs. Dispersion of these particles also develops a random and more heterogeneous plastic flow behaviour within the matrix and promotes fragmentation of the lamellar structures seen in the AA1100 alloy. This behaviour as well as the gradual increase in the percentage of HAGBs and reduction in the grain aspect ratio during the next cycles results in the formation of a more homogenous submicrometre grain structure in the AA3003 alloy (Fig. 6). TEM micrograph depicted in Fig. 8 confirms that the grain size of the sample processed by 10 ARB cycles reaches a nanoscale value. The microstructure mostly consists of grains with a size in the range of 200-300 nm which are surrounded by clear boundaries.

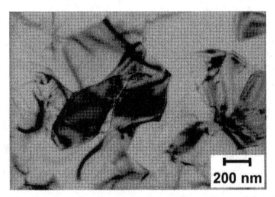

Fig. 8. TEM micrograph showing ultrafine grains in rolling plane of the AA1100 sample after 10 ARB cycles

3.1.2 ASB processing

Optical micrographs of the longitudinal sections of ASB samples are shown in Fig. 9. Previous study on the bond quality of spin-bonding process showed that the threshold thickness reduction for Al/Al bonding by conditions of this work is about 24% (Mohebbi & Akbarzadeh, 2010a). Therefore, it is expected that a good bond is obtained at thickness reduction of 50%. While no interface was observed on the sections after polishing, the bond interfaces are distinguished by etching (Fig. 9). Since subsequent ASB cycles improve the previous bonds, the last bond interface of each specimen is clearer than the previous ones. It is obvious in this figure that the external layer undertakes more thickness reduction than the internal one at each cycle, so that the ratio of external layer thickness to internal one is between 0.40 and 0.45. It should be noted that during the process more elongation were observed in the length of external tube than the internal one, confirming the mentioned thickness strain distribution.

Number of ASB cycles	1	2	3	4
Fraction of the HAGBs	0.54	0.76	0.83	0.84
Grain thickness (nm)	409	306	176	186
Grain length (nm)	885	717	465	419

Table 3. Results of the EBSD analysis

TEM micrographs and the corresponding selected area diffraction (SAD) patterns of the specimens ASBed by one, two and four cycles are shown in Fig. 10. The specimen after one cycle shows a grain size of about 1.5 μm. Large number of subgrain boundaries can be observed in this specimen. The specimen after two cycles illustrates ultra-fine grains with the size of approximately 600 nm. The grain size of about 150 nm is observed in the 4-cycle specimen. The SAD pattern illustrates a more complex shape comparing to the first and second cycles of ASB.

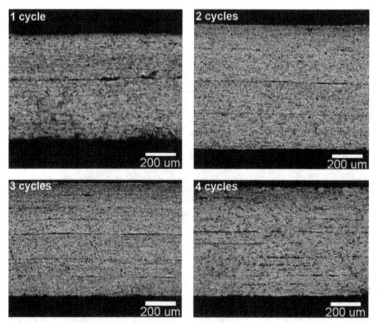

Fig. 9. Optical micrographs obtained from the longitudinal section of the AA1050 specimens after various cycles of ASB

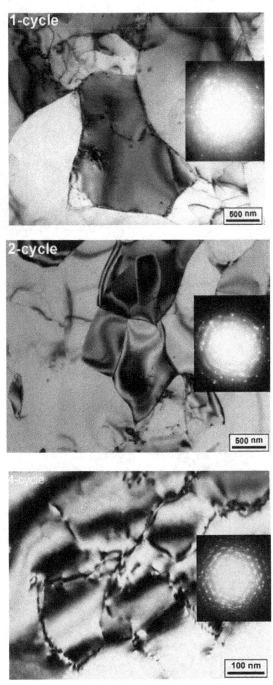

Fig. 10. TEM micrographs and SAD patterns of AA1050 specimens processed by one, two
and 4 cycles of ASB

Fig. 11 illustrates the low angle and high angle grain boundary maps recorded on the radial (r)-longitudinal (z) plane of the ASBed specimens. In this figure, which is obtained from the EBSD analysis, low angle grain boundaries (LAGBs) are depicted by red lines where the misorientation between two points of a step is between 2° and 15° and high angle grain boundaries (HAGBs) with misorientations larger than 15° are drawn in the black lines. Comparing to the very locally analysis of TEM micrographs, this figure clearly shows the

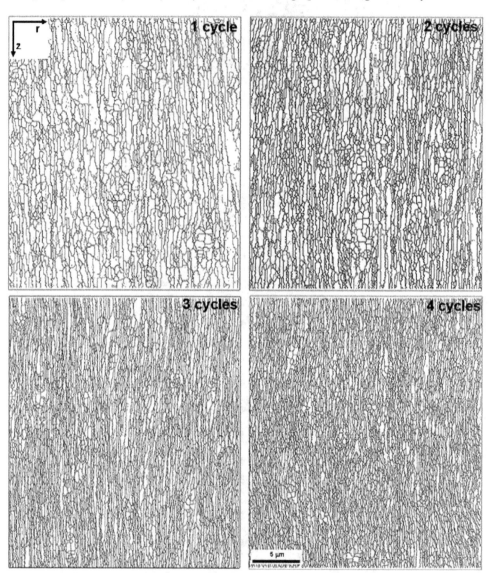

Fig. 11. High angle (black) and low angle (red) grain boundary maps of AA1050 specimens processed by various cycles of ASB

grain refinement and increment of HAGBs fraction by increasing the ASB cycles. Quantitative analysis of the EBSD patterns is presented in Table 3 and Figs. 12 and 13. Regarding the elongated grains, grain size is described by the grain thickness and grain length in radial and longitudinal directions, respectively. This is considered and applied for both subgrain and grain boundaries.

Fig. 12 demonstrates that both grain thicknesses and lengths decrease by increasing the ASB cycles, so that the grain thickness and length are, respectively, 186 and 419 nm after the fourth cycle. In addition to grain refinement, it can be seen in Figs. 11 and 12 that the HAGBs fraction is increased from 54% after the first cycle to 84% after the fourth cycle of ASB. The grain boundary misorientation increments through the ASB cycles are clearer in depiction of misorientation distributions, Fig. 13. The Mackenzie distribution for a random polycrystalline material is shown by black line. It is observed that while the misorientation distribution after the first cycle is considerably far from the random, it becomes much closer to a random distribution during the next ASB cycles.

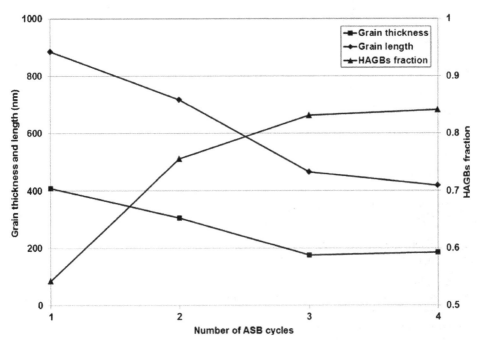

Fig. 12. Evolution of grain thickness and length and fraction of the HAGBs during the ASB cycles obtained from the EBSD analysis

An analytical model was recently developed (Mohebbi & Akbarzadeh, 2010a) to study the strain history of this process. In this model, the radial strain increment at each revolution is calculated. Total strain at each time is then determined by summation of these increments. The strain rate is calculated by dividing these strain increments to the deformation time of each one (Mohebbi & Akbarzadeh, 2010b). It was demonstrated that the material undergoes a small deformation with high strain rate at any exposure to roller, while at the time between the steps, the strain rate downfalls to zero. For thermally activated deformation and restoration processes, which occur even at room temperature by large strains (Estrin et al., 1998), the microstructural evolution and flow stress are dependent on the deformation temperature, strain and strain rate. The strain rate ($\dot{\varepsilon}$) and deformation temperature (T) are included into the Zener–Hollomon Parameter (Z), which is defined as:

$$Z = \dot{\varepsilon} \, \exp{(Q/RT)} \tag{1}$$

where Q is the activation energy for operative process and R is the gas constant (Zener & Hollomon, 1944).

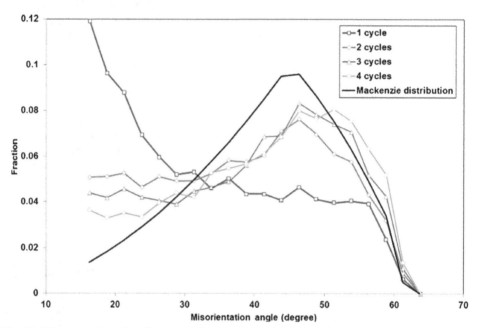

Fig. 13. Misorientation distribution of the specimens processed by various cycles of ASB comparing to the Mackenzie distribution

Generally, by increasing this parameter, subgrain size at dynamic recovery and steady state grain size at dynamic recrystallization are decreased and the flow stress is increased (Humphreys & Hatherly, 2004). Tsuji et al. (2003b) have studied the effect of strain rate on microstructural changes in deformation of the ultra-fine grained (UFG) aluminum produced by ARB. They had expected that higher speed of deformation can result in finer grain subdivision, because larger amount of dislocations would operate at higher strain rates. However, their results showed that grain size rather increases with increasing the strain rate, due to heat of deformation. Therefore, they concluded that higher strain rate and simultaneous cooling are favorable for producing finer grain size in SPD. However, as deformations with strain rates higher than 10 s^{-1} are adiabatic processes (Ryazanov et al., 2003), simultaneous cooling is not effective on deformation temperature and therefore, conclusion of Tsuji et al. (2003b) cannot be correct.

Controlled cooling of the specimen in tensile test maintains the normal temperatures of the tests conducted at 10^{-2} to 10^{-1} s^{-1}. For higher strain rates, the only way is to perform the test by applying strain in increments small enough to determine the negligible temperature rises during the incremental cooling periods (Sevillano et al., 1981). Therefore, it can be said that increasing the strain rate in common SPD processes causes a high temperature rise due to continuous induction of a large strain, which prevents significant promotion of Z. In case of ASB, however, it is possible to increase the strain rate without temperature rise. Deformation with high strain rate and low temperature rise is a very interesting characteristic of ASB process for evaluation of the pure effects of strain rate on the microstructural evolutions and resulted mechanical properties without temperature rise.

This research group's work has shown (Mohebbi & Akbarzadeh, 2010b) that due to large amount of redundant strain and high Zener–Hollomon parameter, as the characteristics of ASB process, the grain refinement and the rate of microstructural evolution are expected to be higher in comparison to other SPD processes. This grain refinement occurred by common mechanism of grain refinement in SPDs of pure aluminum. Increase of dislocation density and its accumulation in cell structure by dynamic recovery leads to formation of subgrains at early stages. This is followed by increase of the misorientations by strain induced transition of low angle to high angle grain boundaries at next stages (Pirgazi et al., 2008b). This mechanism is confirmed by the TEM micrographs and SAD patterns as well as the EBSD analysis. As can be seen in Figs. 11-13, a large fraction of LAGBs is formed after the first cycle. After the fourth cycle, while the grains are more refined, the fraction of HAGBs is considerably increased and the boundary misorientation distribution is close to the random one (Mackenzie distribution).

3.2 Microtexture

Figure 14 shows φ_2 = 45, 65 and 90° sections of the initial orientation distribution functions (ODFs) of both aluminium sheets before the ARB process (fully annealed sheets). It is seen that the initial texture of these samples is mainly composed of a dominant rotated cube {001}<100> which is characteristic of the recrystallisation state. The ODFs corresponding to the samples of Figs. 3 and 4 are illustrated in Figs. 15 and 16 respectively. Only the φ_2 = 45 and 90° sections of ODF are represented because these sections contain the most important texture components for ARB processed aluminium sheets. It is seen that the main texture components in the mid layers of both materials are Copper {1 1 2}<1 1 1> located at φ_1, Φ, φ_2 ≡ 45, 65 and 90° and Dillamore {4 4 11}<11 11 8> at φ_1, Φ, φ_2 ≡ 90, 27 and 45° sections and there is no evidence of shear components. The ODFs of these ARB processed samples were generally similar and the overall texture intensity and the concentration of the contour lines increase continuously with increasing strain. However, it should be noted that in comparison with the AA3003 alloy, a more severe texture was observed in AA1100 alloy during the ARB process and the strength of the overall textures developed at the eighth cycle was 18.4 and 28.2 times the random levels, respectively.

The presence of second phase particles can also change the microtextural evolution in the ARB processed aluminium sheets. The comparison of the ODFs presented in Figs. 15 and 16 indicates that the developed texture in the AA1100 alloy is much stronger than the particle containing alloy. The evolution of a strong texture during the ARB process is attributed to the cyclic nature of this process. Heason and Prangnell (2002a) investigated the texture evolution during the ARB process of the AA1100 aluminium sheets by using a model based on the FC-Taylor approach. According to their model, the surface shear texture which deforms in plane strain compression (in the mid layer) rotates towards the copper orientation during the next cycle. In contrast, this model predicts that all the rolling

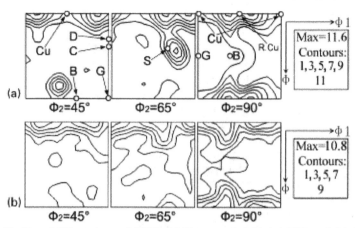

Fig. 14. φ₂=45, 65 and 90° sections of ODF of fully annealed (a) AA3003 and (b) AA1100 aluminium sheets before ARB process: standard texture components are indicated with symbols: Cu: cube; R.Cu: rotated cube; C: copper; B: brass; G: Goss; D: Dillamore

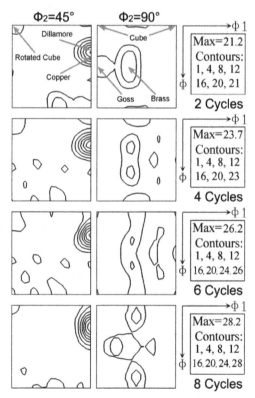

Fig. 15. φ₂=45 and 90° sections of ODFs corresponding to AA1100 samples processed by various ARB cycles: these ODFs were calculated assuming orthorhombic sample symmetry

components (copper, S and brass) rotate towards the shear orientation when they come to the surface of the sheet. This can also explain the increase in the intensity of copper and Dillamore components (which are close to each other) as the main texture components observed in this study. These authors have also reported that this strong texture leads to the development of coarse unrefined bands and prevents the full grain refinement to a submicrometre scale during the ARB process of a single phase aluminium alloy. Whereas, the presence of the second phase particles in the AA3003 alloy results in large local lattice rotation and can cause different textures compared with the rest of the matrix. Thus, the overall texture intensity will be much weaker than the AA1100 alloy.

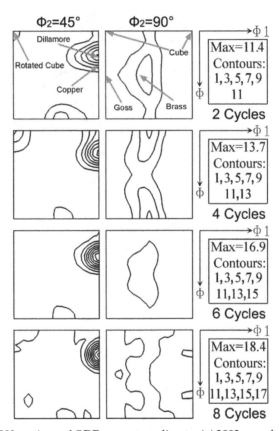

Fig. 16. $\varphi_2=45$ and $90°$ sections of ODFs corresponding to AA3003 samples processed by various ARB cycles

Kim et al. (2005) reported similar results in ARB processed AA8011 aluminium alloy. According to their study, the textural change from a shear texture to a rolling texture at the sheet centre during the ARB process contributed to an increase in the fraction of high angle boundaries. Also, a large number of second phase particles in the AA8011 alloy sheets weakened the texture intensity which is due to the inhomogeneous deformation around the second phase particles. A weak and more spread texture caused by the presence of the second phase particles contributes to the transformation from low angle subgrain boundaries to high angle grain boundaries by increasing the misorientation between two adjacent subgrains and removes the unrefined bands from the microstructure. For a better understanding of the texture evolution, the main FCC fibers were calculated and plotted in Fig. 17. Generally, with increasing the number of cycles, the intensity of the α fiber decreased while the intensities of the β and τ fibers increased. The α fiber, running from Goss to brass in Fig. 17a, indicates how the brass component decreased during the ARB process. It is seen that during all ARB cycles the intensity of Goss component is negligible. Fig. 17b shows the intensity of the β fibers.

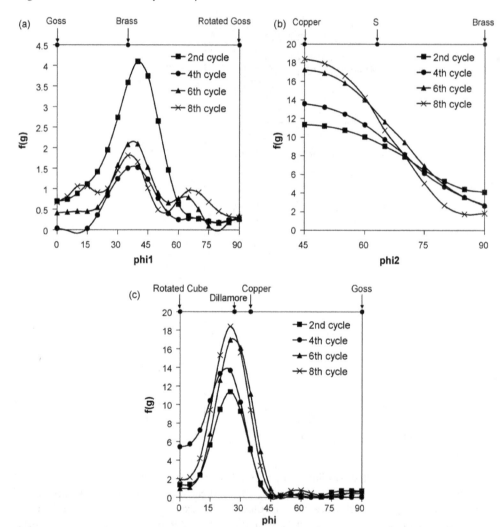

Fig. 17. Intensity of the FCC fibers in the ARB processed aluminum sheets: (a) β-fiber, (b) τ-fiber, and (c) α-fiber

It shows that in the early stages of the ARB process the intensity of the β fiber, running from copper over S to the brass orientation, is almost homogeneous, but with increasing the number of cycles the intensity of the copper component gradually grows at the expense of the brass and S components. The τ fiber is presented in Fig. 17c. It is observed that in this material the texture evolution is mainly centered around the copper and Dillamore components, and with an increasing number of cycles these components further intensify.

3.2.1 Modeling of texture evolution

In order to interpret the texture change involved in the ARB process, the texture development was simulated with the Alamel model (Van Houtte et al., 2005, 2006). The deformation during ARB processing is approximated by a two-dimensional velocity gradient tensor which is composed of a compressive strain (ε_{zz}, $\varepsilon_{yy} = 0$ and $\varepsilon_{zz} = -\varepsilon_{xx}$) added with a simple shear strain (ε_{xz}). The texture simulations have been performed on the initial texture of Fig. 14 for different strain modes with various ratios of the simple shear component (ε_{xz}) to the plane strain compression components. The imposed shear during the ARB process is characterized by the shear coefficient K ($K = \varepsilon_{xz}/\varepsilon_{zz}$).

The texture simulation for the subsurface region has been performed with a large value of the shear coefficient (K= 2.4) due to localization of shear deformation in this region. The texture prediction for the mid-thickness part has been performed under the deformation condition with K= 0.1. It is assumed here that even in the middle layers of the sheet there will still be a small amount of frictional shear as the ARB experiment was carried out in totally dry conditions (i.e. without lubrication). Due to the specific stack-and-roll geometry of the ARB process, the subsurface texture after the first ARB pass appears in the central layer of the composite sample in the second ARB pass. Hence, in order to simulate the texture evolution in the middle layer in the ARB pass n_i the Alamel model was applied on the subsurface from previous pass $n_{(i-1)}$ with a deformation mode that was characterized by K= 0.1. Fig. 18 presents results of the texture predictions for second, fourth, sixth and eighth ARB cycles. The calculated textures are in very good qualitative and quantitative agreement with the experimentally measured ODFs (Figs. 16 and 18).

The model calculation has produced the textures which exhibit a strong rolling β-fiber. It is important to notice, though, that the Dillamore and S components are far more important than the brass orientation, which corresponds very well to the experimentally observed texture, cf. Fig. 16. Furthermore, intensities of both S and brass orientations increase insignificantly whereas a considerable strengthening of the Dillamore orientation is observed during ARB process. Hence, it is shown here, on the basis of crystal plasticity modeling, that the sequence of rolling, cutting and stacking, which is characteristic for the ARB process eventually triggers a mechanism which leads to strengthening of the Dillamore component in the middle layers of the ARB sample.

Rotation of different texture components during ARB cycles, i.e. shear to copper and Dillamore components in the midsection under the plane strain compression mode and copper, S and brass to the shear component in the surface layer due to the shear deformation, is the unique feature of ARBed aluminum sheets (Heason & Prangnell, 2002a; Kim et al., 2005). This can be considered as the result of introduction of surface layer to the midsection during the next cycle and it is also associated with the increase in the number of interfaces. In fact, these interfaces induce additional strain during rolling and are responsible for the unique property.

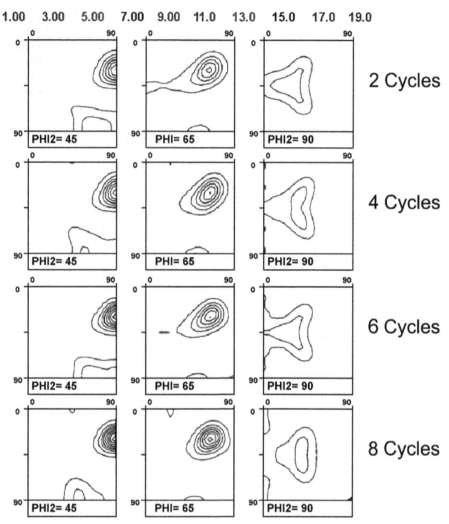

Fig. 18. Simulation of the texture development in the center region during different ARB cycles

3.3 Mechanical properties
3.3.1 ARB samples

Second phase particles in AA3003 alloy can also accelerate the rate of dynamic recovery which is very likely to occur in warm rolled aluminum sheets. Accumulation of dislocations in the vicinity of these particles leads to rapid transformation of low angle sub-grain boundaries to high angle grain boundaries and decreases the dislocation density inside the grains. Static recovery during pre-heating may also contribute to the reduction of dislocation density and promote the softening behavior. The variation of the microhardness with increasing strain during the ARB process confirms the occurrence of recovery in this phenomenon, Fig. 19. The hardness curve in this figure corresponds well with the microstructural changes mentioned above. Initial work hardening caused by the reduction of grain size and the growth of dislocation density inside the crystalline lattice leads to a rapid increase in microhardness at the first ARB cycle (ε_{vM} = 0.8). During the following cycles, static or dynamic recovery caused by the interpass annealing treatment, the heat of deformation and the accumulation of dislocations in the vicinity of the second phase particles, prevents a further increase of the hardness and a plateau around 80 HV is established.

An interesting feature of this figure is the significant increase of the hardness value (close to two times of the initial value) after the eighth cycle, while the sample exhibits a fully recovered microstructure and probably still provides an acceptable ductility. According to the results presented here, different mechanisms of grain refinement can be attributed to the different levels of strain. During the first two cycles of ARB (ε < 1.6) grain subdivision is the dominant mechanism. This mechanism promotes the formation of a fibrous microstructure of elongated grains.

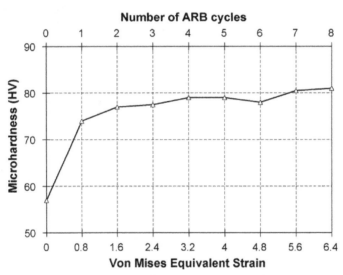

Fig. 19. Variation of Vickers microhardness of AA3003 alloy with increasing strain during ARB process

The results of the mechanical properties clarified that the strength of the ARBed aluminum sheets increases with increasing the number of cycles, Fig. 20. In order to study the relation between the mechanical properties and the microstructural changes, the tensile strength and the 0.2% yield strength were plotted as a function of minus square root of grain size in Fig. 21. The strength of the ARBed aluminum sheets is in a good conformity with Hall–Petch relationship, even for submicron grain structures. It was shown (Pirgazi et al., 2008b) shows that the amount of work hardening after yielding is rather small and necking occurs suddenly and causes the limited uniform elongation. It should be noted that the described microstructural evolution corresponds to the changes in mechanical properties very well. This suggests that the strength of the ARB samples is mainly attributed to grain boundary strengthening, though the details of the strengthening mechanism in the SPD materials should be discussed more. As has been already known, the ARB processed specimens have the dislocation substructures inside the UFGs as well as the elongated grain morphology. This means that the strength of these specimens might be affected not only by grain refinement strengthening but also by strain hardening. It has been proposed that when a deformation structure consists of a mixture of low and high angle grain boundaries, the strength can be considered as the sum of dislocation strengthening from LAGBs and grain size strengthening from HAGBs (Hansen, 2004). However, the detailed discussion about the Hall–Petch relationship and strengthening mechanisms should be done in the future works. In a similar study by Park et al. (2001) on AA6061 alloy processed by the ARB at 250 °C it has been reported that the rapid increase in strength at relatively low strains is mainly due to the work hardening caused by an increase in dislocation density and formation of the sub-grains.

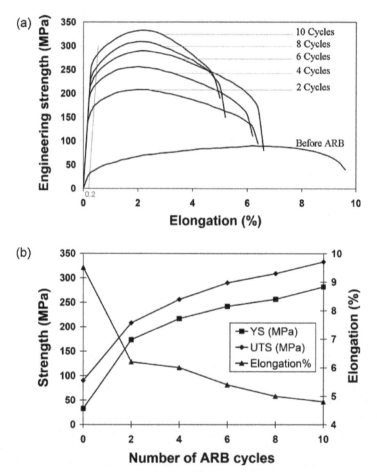

Fig. 20. (a) Engineering stress–strain curves and (b) the corresponding tensile properties of AA1100 aluminum sheets processed at different ARB cycles

During the following cycles, in which the incremental increase of the strength becomes smaller, the dislocation density is almost constant (Parh et al., 2001; Costa et al., 2005). The low density of dislocations at relatively high strains is due to the dynamic recovery (Costa et al., 2005), or absorption of dislocations into the grain boundaries (Prangnell et al., 2001). In any way, the increase of the volume fraction of ultra-fine elongated grains due to increase in the fraction of HAGBs can be contributed to the strengthening in the medium levels of strain $(1.6 < \varepsilon < 4.8)$. Finally, the large strength at relatively high strains $(\varepsilon > 4.8)$ can be explained by the development of ultra-fine grained structures.

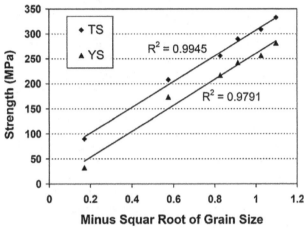

Fig. 21. Strength of ARBed AA1100 aluminum sheets versus the grain size. The grain sizes reported in this figure are the size of grains bounded by high angle grain boundaries

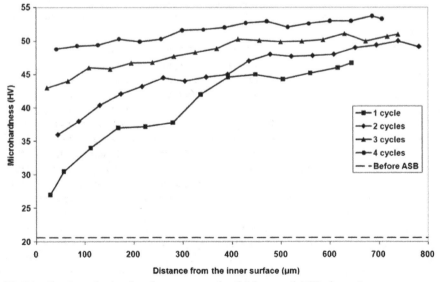

Fig. 22. Distribution of microhardness across the thickness of ASBed specimens

3.3.2 ASB samples

Distributions of microhardness across the thickness of specimens are compared in Fig. 22. The hardness of annealed specimen before ASB is shown by a dash line. It can be seen in this figure that the hardness is increased from the inner layers to the outer ones. Two trends in the hardness evolutions during the ASB cycles can be derived from this figure. First, the degree of hardness is increased, and second, the hardness distribution across the thickness of the tubes is homogenized. Fig. 23 shows the tensile stress–strain curves of the ASBed tubes.

Tensile properties derived from these curves are summarized in Fig. 24. The yield strength (at 0.2% offset strain) and the tensile strength are increased by the ASB cycles, although the growth rate is diminished. So that, the yield strength is increased up to 194 MPa after four cycles of ASB, which is about five times higher comparing to the initial material (40 MPa). Fig. 24 also shows that the difference between the yield and the tensile strengths is slightly decreased during the ASB cycles. Regarding the rupture elongation, a sharp drop is observed after the first cycle from 38% to 9%. Decrease of the elongation at subsequent cycles is quite low as compared to that of the first cycle. Hardness distribution across the thickness of the tubes confirms that materials below the outer surface undergo heavier plastic strain in comparison to the internal ones. This is due to not only higher thickness reduction but also higher redundant strains at the outer regions (Mohebbi & Akbarzadeh, 2010c).

Although this inhomogeneous hardness distribution is observed at all specimens, its intensity is decreased by increasing the number of ASB cycles. Two reasons can be mentioned for this scheme of hardness evolution. First, at each cycle of ASB, a tube is used as internal tube and so, its severely strained external layer is located within the interior regions. By this periodical entrance of the highly deformed outer materials within the thickness, the hardness distribution becomes more homogeneous by increasing the number of ASB cycles. Second, due to saturation of dislocation density and microstructural evolutions, the hardness and strength are always saturated at large strains (Valiev et al., 2000). Therefore, despite of inhomogeneous deformation, hardness becomes homogenized at large strains.

As it can be seen in Figs. 23 and 24, the yield strength and the tensile strength of the specimens are increased by increasing the ASB cycles, although its rate is gradually diminished. Grain refinement by mentioned mechanism, i.e. increase of dislocation density and formation of subgrains at early stages as well as increase of the misorientations at next stages are responsible for this strengthening (Pirgazi et al., 2008b; Sevillano et al., 1981). While based on this mechanism a steady state plateau is expected for strength of material, Fig. 24 shows that yield and tensile strengths are continuously increased. One may conclude that the strain induced by four cycles is not enough for strength saturation, especially at high value of Z in this work. However, it is shown that this continuous increase of the strength can be attributed to the through thickness inhomogeneity of hardness. Considering the mentioned hardness evolutions, it can be said that outer materials are severely work hardened after the two first cycles in comparison to inner ones. Therefore, in spite of strength saturation of the outer material, inner material has the possibility of work hardening at following cycles. This work hardening has two effects on the tensile properties. The first effect is continuous increase of the strength. By increasing the ASB cycles, inner materials are work hardened and therefore, the strength is increased. The second effect is related to the ratio of tensile to yield strength in tensile tests. As a result of work hardening of the inner materials, the overall flow strength of the specimen is increased during tension. That is why there is a high difference between the yield strength and the tensile strength in this work in comparison to that of ARB (Pirgazi et al., 2008b). By this explanation, more hardness inhomogeneity causes higher ratio of tensile to yield strength. Therefore, decrease of the difference between the yield and tensile strengths by the ASB cycles in Fig. 24, is mostly related to the hardness homogenization.

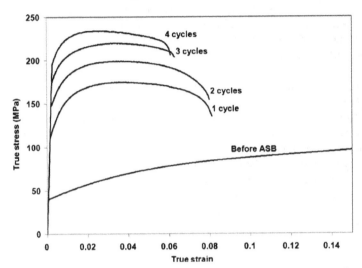

Fig. 23. Tensile stress–strain curves of the tubes processed by various ASB cycles

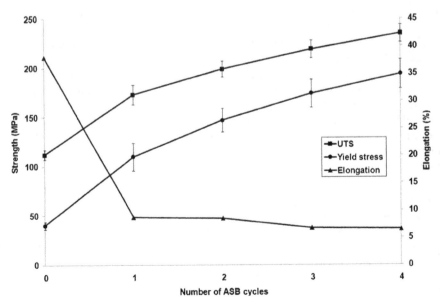

Fig. 24. Development of the tensile properties of tubes processed by various ASB cycles

Fig. 24 demonstrates a severe drop in elongation after the first cycle of ASB while its rate of reduction is very low at next cycles. This result is similar to the reported ones for ARB (Lee et al., 2002; Pirgazi et al., 2008b). Uniform elongation, which is reduced by increasing the ASB cycles, is related to onset of the plastic instability or necking. Plastic instability, on the other hand, is a function of work hardening and consequently, of the ratio of tensile to yield strength (Semiatin & Jonas, 1984). Therefore, it can be said that since this ratio is decreased by increasing the ASB cycles, the uniform elongation is decreased. In other words, inner materials of specimens after the early cycles of ASB, whose work hardening is not saturated, are work hardened during the tensile test, leading to delay in necking. At next cycles, due to saturation of work hardening across the entire thickness, plastic instability starts earlier and causes a less uniform elongation.

It is seen in Fig. 23 that post-uniform elongation includes a considerable portion of the total elongation. Previous studies on SPD have demonstrated that strain rate sensitivity of fcc metals increases by decreasing the grain size (Wang & Ma, 2004; Wei et al., 2004). High strain rate sensitivity, on the other hand, causes the increase of both uniform and post uniform elongations (Semiatin & Jonas, 1984). Takata et al. (2008) have illustrated that uniform elongation of commercially pure aluminum increases with increasing the strain rate in the specimens with a grain size larger than 1 µm, while post-uniform elongation increases with increasing the strain rate in the sub-micrometer grain size specimens. Therefore, it can be concluded that high strain rate sensitivity is effective on post-uniform elongation of aluminum with sub-micron grains. In fact, the high magnitude of post-uniform elongation in Fig. 23 can be mostly related to the high strain rate sensitivity.

4. Conclusions

EBSD analysis confirmed that the development of UFGs by ARB process is carried out via several mechanisms at different levels of strain. Grain subdivision as well as development of sub-grains is the major mechanism at the early stages of ARB. Strain induced transition of low angle to high angle grain boundaries and the formation of a thin lamellar structure occurs at the medium levels of strain. Finally, the progressive fragmentation of these thin lamellar structures into more equi-axed grains is the dominant mechanism at relatively high strains which leads to reduction of grain size to less than 500 nm. The presence of second phase particles in AA3003 aluminum sheets promotes the grain refinements and accelerates the occurrence of dynamic recovery. With an increasing number of cycles, the overall texture intensity increases and a very strong texture is developed which exhibits a limited number of sharp components. Large lattice rotation around these particles leads to the increase of local misorientation, evolution of a much weaker texture and development of a more homogeneous submicrometre grain structure in the AA3003 alloy. The Dillamore and the copper components are the main texture components of this material and they are sharpened with increasing the number of cycles. This texture evolution could be modeled with the Alamel crystal plasticity model and taking into account the specific geometry of the ARB process.

Generally, with increasing the number of cycles, the intensity of the α fiber decreased while the intensities of the β and τ fibers increased. It is shown that in the early stages of the ARB process the intensity of the β fiber, running from copper over S to the brass orientation, is almost homogeneous, but with increasing the number of cycles the intensity of the copper component gradually grows at the expense of the brass and S components.

It is demonstrated that due to incremental deformation in ASB process, high value of strain rate without considerable temperature rise is applied, leading to a high degree of Zener–Hollomon parameter, as a characteristic of this SPD process. The grain structure of commercially pure aluminum is refined by this process and an average grain thickness and length of 186 and 419 nm are developed after the four cycles of ASB, respectively. TEM micrographs and SAD patterns as well as the EBSD analysis indicate that this grain refinement occurred by formation of subgrains at early stages of severe deformation followed by increase of the misorientations by transition of low angle to high angle grain boundaries at the next stages.

The characterization of mechanical properties revealed that the strength of the aluminum sheets considerably increased by the first two ARB cycles which is attributed to work hardening caused by increase in dislocation density and formation of subgrains. It was also included that the strength of ARBed aluminum sheets obeys the Hall–Petch relationship and corresponds well with microstructural variations. Microhardness distributions across the thickness of the tubes illustrate that because of high equivalent strain at outer regions, the hardness is increased from the inner regions to the outer ones. Due to periodical entrance of the external layer within the thickness and the consequent hardness saturation, the hardness and its homogeneity is increased with increase of the ASB cycles. The yield and tensile strengths of the material are significantly increased up to the values of 194 and 235 MPa, respectively. The scheme of hardness development leads to decrease of the ratio of tensile strength to yield strength and consequently to decrease of the uniform elongation.

5. Acknowledgments

Professors Leo Kestens and Roumen Petrov's permission for access to EBSD analysis and their help in the EBSD mapping and discussion about the texture analysis, at Ghent University in Belgium, are gratefully acknowledged. The author would like to express his sincere gratitude to Mr. Hadi Pirgazi and Mr. M.S. Mohebbi for their excellent research works at Sharif University led to this manuscript. The support of research office of Sharif University of technology is also acknowledged.

6. References

Berski, S., Dyja, H., Maranda, A., Nowaczewski, J. & Banaszek, G. (2006). *Journal of Materials Processing Technology*, Vol. 177, pp. 582–586

Chen, Z., Ikeda, K., Murakami, T, Takeda, T., Xie, J.X. (2003). *Journal of Materials Processing Technology*, Vol. 137, pp. 10–16

Chowdhury, S.G., Srivastava, V.C., Ravikumar, B. & Soren, S. (2006a). *Scripta Materialia*, Vol. 54, pp. 1691–1696

Chowdhury, S.G., Dutta, A., Ravikumar, B. & Kumar, A. (2006b). *Materials Science and Engineering A*, Vol. 428, pp. 351–357

Costa, A.L.M., Reis, A.C.C., Kestens, L. & Andrade, M.S. (2005). *Materials Science and Engineering A*, Vol. 406, pp. 279–285

del Valle, J.A., Pérez-Prado, M.T. & Ruano, O.A. (2005). *Materials Science and Engineering A*, Vols. 410/411, pp. 353–357
Estrin, Y., Toth, L.S., Molinari, A. & Bréchet, Y. (1998). *Acta Materialia*, Vol. 46, pp. 5509–5522
Hansen, N. (2004). *Scripta Materialia*, Vol. 51, pp. 801–806
Heason, C.P. & Prangnell, P.B. (2002a). *Materials Science Forum*, Vols. 408–412, pp. 733– 738
Heason, C.P. & Prangnell, P.B. (2002b). *Proc. of Conference on Nanomaterials by Severe Plastic Deformation-NANOSPD2,*, Dec. 9-13, Vienna, Austria
Horita, Z., Smith, D.J., Furukawa, M., Nemoto, M., Valiev, R.Z. & Langdon, T.G. (1996). *Journal of Material Research*, Vol. 11, pp. 1880–1890
Huang, X., Tsuji, N., Hansen, N. & Minamino, Y. (2003). *Materials Science and Engineering A*, Vol. 340, pp. 265–271
Humphreys, F.J. & Hatherly, M. (2004). *Recrystallization and Related Annealing Phenomena*, Elsevier, Oxford
Kim, H.W., Kang, S.B., Xing, Z.P., Tsuji, N. & Minamino, Y. (2002). *Materials Science Forum*, Vols. 408–412, pp. 727–732
Kim, H.W., Kang, S.B., Tsuji, N. & Minamino, Y. (2005). *Metallurgical and Materials Transaction A*, Vol. 36, pp. 3151–3163
Kim, Y.S., Kang, S.H., & Shin, D.H. (2006). *Materials Science Forum*, Vols. 503/504, pp. 681– 686
Kolahi, A., Akbarzadeh A. & Barnett, M.R. (2009). *Journal of Materials Processing Technology*, Vol. 209, pp. 1436-1444
Lee, S.H., Saito, Y., Sakai, T. & Utsunomiya, H. (2002). *Materials Science and Engineering A*, Vol. 325, pp. 228–235
Lee, S.H., Lee, C.H. & Lim, C.Y. (2004). *Materials Science Forum*, Vols. 449–452, pp. 161–164
Min, G., Lee, J.M., Kang, S.B. & Kim, H.W. (2006). *Materials. Letters*, Vol. 60, pp. 3255–3259
Mohebbi, M.S. (2009a). M.Sc. Thesis, Sharif University of Technology, Tehran, I.R. Iran
Mohebbi, M.S. & Akbarzadeh, A. (2009b). *Proc. of the 2nd Int. Conference on Ultrafine Grained and Nanostructured Materials (UFGNS2009)*, Nov. 14-15, Tehran University, Tehran, I.R. Iran
Mohebbi, M.S. & Akbarzadeh, A. (2010a). *Journal of Materials Processing Technology*, Vol. 210, pp. 510–517
Mohebbi, M.S. & Akbarzadeh, A. (2010b). *Materials Science and Engineering A*, Vol. 528, pp. 180–188
Mohebbi, M.S. & Akbarzadeh, A. (2010c). *Journal of Materials Processing Technology*, Vol. 210, pp. 389–395
Park, K.T., Kwon, H.J., Kim, W.J., & Kim, Y.S. (2001). *Materials Science Engineering A*, Vol. 316, pp. 145–152
Pérez-Prado, M.T., del Valle, J.A. & Ruano, O.A. (2004). *Scripta Materialia*, Vol. 51, pp. 1093– 1097
Pirgazi H., Akbarzadeh, A., Petrov, R., Sidor, J. & Kestens, L. (2008a). *Materials Science and Engineering A*, Vol. 492, pp. 110–117
Pirgazi, H., Akbarzadeh, A., Petrov, R. & Kestens, L. (2008b). *Materials Science and Engineering A*, Vol. 497, pp. 132–138

Pirgazi, H. & Akbarzadeh, A. (2008c). *Proc. of the 2nd Conference on Nanostructures (NS2008)*, March 11-14, Kish University, Kish Island, I.R. Iran

Pirgazi, H. & Akbarzadeh, A. (2009). *Materials Science and Technology*, ol. 25, No. 5, pp. 625-631

Prangnell, P.B., Bowen, J.R. & Gholinia, A. (2001). *Proceedings of the 22nd Risø´ International Symposium on Materials Science*, pp. 105-126

Reis, A.C.C. & Kestens, L. (2005). *Solid State Phenomena*, Vol. 105, pp. 233-238

Reis, A.C.C. , Kestens, L. & Houbaert, Y. (2005). *Materials Science Forum*, Vols. 495-497, pp. 351-356

Richert, J. & Richert, M. (1986), *Aluminum*, Vol. 62, pp. 604-607

Rios, P. R. & Padilha, A. F. (2003). *Materials Research*, Vol. 6, No. 4, pp. 605-613

Ryazanov, A.I., Pavlov, S.A. & Kiritani, M. (2003). *Materials Science and Engineering A*, Vol. 350, pp. 245-250

Saito, Y., Utsunomiya, H., Tsuji, N. & Sakai, T (1999). *Acta Materialia*, Vol. 47, No. 2, pp. 579-583

Saito, Y., Tsuji, N., Utsunomiya, H., Sakai, T. & Hong, R.G. (1998). *Scripta Materialia*, Vol. 39, No. 9, pp. 1221-1227

Semiatin, S.L. & Jonas, J.J. (1984). *Formability and Workability of Metals*, American Society for Metals, Ohio, USA

Sevillano, J.G., Van Houtte, P. & Aernoudt, E. (1981). *Progress in Materials Science*, Vol. 25, pp. 69-134

Sherby, O.D. and Wadsworth, J. (2001). *Journal of Materials Processing Technology*, Vol. 117, pp. 347-353

Sponseller, D.L., Timmons, G.A. & Bakker, W. T. (1998). *Journal of Materials Engineering and Performance*, Vol. 7, No. 2, pp. 227-238

Takata, N., Okitsu, Y. & Tsuji, N. (2008). *J. Materials Science*, Vol. 43, pp. 7385-7390

Tsuji, N., Ueji, R. & Minamino, Y. (2002a). *Scripta Materialia*, Vol 47, pp. 69-76

Tsuji, N., Ito, Y., Saito, Y. & Minamino, Y. (2002b). *Scripta Materialia*, Vol. 47, pp. 893-899

Tsuji, N., Saito, Y., Lee, S.H. & Minamino, Y. (2003a). *Advanced Engineering Materials*, Vol. 5, No. 5, pp. 338-344

Tsuji, N., Toyoda, T., Minamino, Y., Koizumi, Y., Yamane, T., Komatsu, M. & Kiritani, M. (2003b). *Materials Science and Engineering A*, Vol. 350, pp. 108-116

Valiev, R.Z., Krasilnikov, N.A. & Tsenev, N.K. (1991). *Materials Science Forum Engineering A*, Vol. 137, pp. 35-41

Valiev, R.Z., Islamgaliev, R.K. & Alexandrov, I.V. (2000). *Progress in Materials Science*, Vol. 45, pp. 103-189

Van Houtte, P., Li, S.Y., Seefeldt, M. & Delannay, L. (2005). *International Journal of Plasticity*, Vol. 21, pp. 589-624

Van Houtte, P., Kanjarla, A.K., Van Bael, A., Seefeldt, M. & Delannay, L. (2006). *European Journal of Mechanics A-Solids*, Vol. 25, pp. 634-648

Wang, X., Li, P. & Wang, R. (2005). *Int. J. of Machine Tools Manufacture*, Vol. 45, pp. 373-378

Wang, Y.M. & Ma, E. (2004). *Materials Science and Engineering A*, Vols. 375-377, pp. 46-52

Wei, Q., Cheng, S., Ramesh, K.T. & Ma, E. (2004). *Materials Science and Engineering A*, Vol. 381, pp. 71-79

Wong, C.C., Dean, T.A. & Lin, J. (2003). *Int. J. of Machine Tools Manufacture*, Vol. 43, pp. 1419–1435

Zener, C. & Hollomon, J.H. (1944). J. Applied Physics, Vol. 15, pp. 22–32

Zhan, Z., He, Y., Wang, D. & Gao, W. (2006). *Surface and Coatings Technology*, Vol. 201, pp. 2684–2689

8

Statistical Tests Based on the Geometry of Second Phase Particles

Viktor Beneš[1], Lev Klebanov[1], Radka Lechnerová[2] and Peter Sláma[3]
[1]*Charles University in Prague, Faculty of Mathematics and Physics, Department of Probability and Mathematical Statistics*
[2]*Private College of Economic Studies, Ltd., Prague*
[3]*COMTES FHT a.s., Metallography, Dobřany*
Czech Republic

1. Introduction

The actual trends in new material development impose the necessity of a thorough knowledge of the relationships between properties and microstructure. In order to meet the requirements on the performance of some materials in their applications, that are often very sophisticated, a very fine tuning of the manufacturing process and its parameters is needed. Therefore, it is indispensable to be able to distinguish between small differences in microstructures produced by the different variations of processing parameters. Another important field of material characterization is the description of microstructure heterogeneities. Such heterogeneities are often related to risks of premature damage nucleation and preferential defects (void, cracks, corrosion, etc.) occurrence and propagation.

One of the most important elements of the microstructure of metallic materials is the set of second phase particles. Particle size and shape distributions and the type of spatial dispersion (homogeneous, long-range or short range ordered, clustered, etc.) are often the major attributes of a particular microstructure (Humphreys & Hatherly, 2004; Polmear, 2006). Thin foils made from aluminium-manganese based alloys, such as AA3003, are the material most frequently used as fins in automotive heat exchangers (Hirsch, 2006). This application imposes very strict requirements on properties and related foil microstructures. The development of an appropriate production technology is contingent on the perfect knowledge of the impact of processing parameters on microstructure transformation, including the changes of the set of particles (Hirsch, 2006; Slámová et al., 2006).

In statistical setting, we deal with microstructures containing random objects in a space or plane, which may be second phase particles, pores, grains and their sections or projections. The question frequently asked is whether two microstructures come from a material with the same geometrical characteristics of microstructure. This statement forms a null hypothesis H_0 and the aim is to develop a statistical two-sample test of H_0 against an alternative hypothesis that the geometrical characteristics are different. In the literature, parametric models of microstructures as random sets are mostly used (Derr & Ji, 2000; Ohser & Mücklich, 2000) and the authors recommend Monte Carlo testing which is based on the possibility of simulating a random set under the null hypothesis. The evaluation of the test is based on a comparison of the test statistics (describing some characteristics (Tewari & Gokhale, 2006a;b) of the random set) obtained from simulated models with that obtained from the microstructure. In the present paper, we apply a nonparametric approach. We will consider some geometrical characteristics, which can be measured by means of image analysis or estimated from the observation in a window. Thus, we obtain a vector (typically of a large dimension) of data, which does not form a random sample since the vector components may be stochastically dependent. For statistical testing, such a dependency is a serious problem. For the data from a single window, such a problem should be considered (Kupczyk, 2006). Under the assumption that we can observe the microstructure in a few independent windows, we put all the data from the individual windows and observations within windows together. This global information for both sets can be compared using $\mathfrak{N}$-distances from probability theory, see (Klebanov, 2005). The same holds in the functional data approach, where data is first transformed to a function. When the number of independent windows is large, the test of H_0 can be transformed into a univariate distribution-free two-sample test. For a smaller number of measurements, the permutation test for H_0 based on $\mathfrak{N}$-distances can be used. In the paper, we consider microstructures with dispersed particles. A statistical method is developed independently of whether we deal with 2D or 3D data. Therefore, in the application presented from metallography, we will call the feature particles, which in fact are 2D particle sections. We distinguish two basic sets of particle characteristics, namely the individual particle geometry and the spatial distribution of particles. Even if these are rather different descriptive tools, we try to construct the test so that it can be applied to both of them in an analogous way. We describe a variety of methods in Section 2 including also a functional data analysis. In the applied part of the paper in Section 3, we present data from thin foils demonstrating the use of the test for a comparison of metallographic samples of aluminium alloys, observed by means of optical microscopy. The numerical results are in Section 4. Finally, in Section 5 we present a general discussion of the methods and interpretations of results. The theoretical statistical background is given in the Appendix.

2. Methods

Several statistical methods for the testing of differences between microstructures containing particles are suggested in this Section. In Subsections 2.1 and 2.2 we describe particle systems by a vector of parameters and the testing is reduced to a test of the null hypothesis that the vector of these random parameters has the same distribution for two random sets A and B. In Subsection 2.3 the functional data approach is used, we compare functions which fit the observed data. Finally in Subsection 2.4 a simulation study concerning the power of the test based on $\mathfrak{N}$-distances is performed in order to see its behavior with respect to different alternative hypotheses.

2.1 Vector approach – individual particle parameters

Here we describe method (I). Generally the individual particles observed in a window (metallographic sample) are not independent. We will assume that n windows of the same size and magnification are observed for two microstructures A and B. Let the windows be each sufficiently far from the other or taken from independent samples so that we can assume independence among windows. We measure the same number k of particles from each window. In our application, the image analyzer scans particles in a window in a meandering way so that we obtain a representative set of particles by taking the first k particles measured from each window. Assume that m geometrical parameters are measured for each particle (corresponding to microstructures A, B). Two independent samples $X_1, \ldots, X_n$ from A and

$Y_1, \ldots, Y_n$ from B with matrices of size $k \times m$ are thus obtained. Now we transform each matrix X_i, Y_i to a vector $\tilde{x}, \tilde{y}$, respectively, of size km:

$$\tilde{x} = (x_{1,1}, \ldots, x_{k,1}, x_{1,2}, \ldots, x_{k,2}, \ldots, x_{1,m}, \ldots, x_{k,m}), \tag{1}$$

$$\tilde{y} = (y_{1,1}, \ldots, y_{k,1}, y_{1,2}, \ldots, y_{k,2}, \ldots, y_{1,m}, \ldots, y_{k,m}),$$

and evaluate an empirical counterpart of the $\mathfrak{N}$-distance (11):

$$\widehat{\mathfrak{N}} = \frac{1}{n} \left[\sum_{i=1}^{n} \sum_{j=1}^{n} \left(2\mathcal{L}(\tilde{x}_i, \tilde{y}_j) - \mathcal{L}(\tilde{x}_i, \tilde{x}_j) - \mathcal{L}(\tilde{y}_i, \tilde{y}_j) \right) \right]^{\frac{1}{2}}, \tag{2}$$

where e.g.

$$\mathcal{L}(s, t) = \|s - t\| \tag{3}$$

is the Euclidean distance between vectors s, t, which is a strongly negative definite kernel, cf. Appendix.

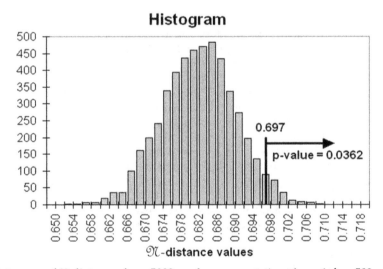

Fig. 1. Histogram of $\mathfrak{N}$-distances from 5000 random permutations ($m = 1, k = 500, n = 20$). The value 0.697 corresponds to the non-permuted case and the p-value is the probability (relative frequency) of $\mathfrak{N}$-distance being larger than this value.

We describe the permutation test (Lehmann & Romano, 2005) of the null hypothesis in more detail, which is used when n is small. Consider K random permutations of $1, \ldots, 2n$. Apply each permutation to long vector $(\tilde{x}_1, \ldots, \tilde{x}_n, \tilde{y}_1, \ldots, \tilde{y}_n)$, and then split the permuted set to the first n and last n vectors and evaluate (2) to obtain K empirical $\mathfrak{N}$-distances (K is recommended to be about 1000). Under the null hypothesis, permutations do not modify the distribution of the random variable $\mathfrak{N}$. From the histogram of these distances including the non-permuted case, we obtain the p-value for the test, which is the probability (under the validity of the null hypothesis) that the random $\mathfrak{N}$-distance is larger than its measured value. A typical example of the test is in Fig. 1, here we reject H_0 since the p-value is smaller than 0.05. If the p-value were greater than 0.05, H_0 would not be rejected. This rule we use in all tests throughout the whole chapter.

We recommend also a simpler test in which the samples are split randomly into three sub samples $\tilde{x}, \tilde{x}', \tilde{x}''(\tilde{y}, \tilde{y}', \tilde{y}'')$, respectively, of size n/3 (assuming it is an integer). Then put (12)

$$U_i = \mathcal{L}(\tilde{x}_i, \tilde{y}_i) - \mathcal{L}(\tilde{x}_i, \tilde{x}_i') \tag{4}$$
$$V_i = \mathcal{L}(\tilde{y}_i', \tilde{y}_i'') - \mathcal{L}(\tilde{x}_i'', \tilde{y}_i'')$$

$i = 1, \ldots, n/3$. The null hypothesis is now equivalent to the hypothesis that U_i, V_i come from the same distribution, which can be tested by an arbitrary univariate two-sample test, e.g. a Kolmogorov-Smirnov test (using STATS package in R language, (Ihaka & Gentleman, 1996)), whose statistic has the form

$$\max_x |H_n^U(x) - H_n^V(x)|,$$

where $H_n^U(x)$ and $H_n^V(x)$ are empirical distribution functions of $U_1, \ldots, U_n$ and $V_1, \ldots, V_n$ correspondingly.

For small values of n, i.e. $n < 120$, however, the loss of information when splitting the files to a size $n/3$ leads to the situation where the use of the asymptotic statistics for the Kolmogorov-Smirnov test is not recommended (Buening & Trenkler, 1978, p.135). The test based on splitting is distribution-free (independent of the underlying distribution of observations), but it has smaller power than the permutation test, see (Klebanov, 2005). Namely, based on simulated samples from multivariate normal distributions and location alternative, it was shown that splitting test has about the same power as permutation test, but based on three times smaller sample size. For the one-dimensional case and the samples from normal distribution and location alternatives, the permutation $\mathfrak{N}$-test has the power very closed to optimal t-test. However, for the samples from the mixture of normal distribution $\mathfrak{N}$-test may be more powerful than t-test. In all situations permutation $\mathfrak{N}$-test is more powerful than Kolmogorov-Smirnov test.

2.2 Vector approach – the spatial distribution of particles

Here we do not evaluate the measurement directly but first the measured information is transformed.

To test the difference in spatial distribution of particles we use m mutual characteristics of particle centroids, among them:

a) a distribution function of the nearest neighbour distance (G-function) (Tewari & Gokhale, 2006b),

b) a contact distribution function (F-function) (Tewari & Gokhale, 2006a),

c) a pair correlation function (pcf) (Ohser & Mücklich, 2000).

Concerning the spatial distribution, we distinguish complete independence (CI), attraction (clustering) and repulsion (regularity). Functions F and G coincide when CI takes place, for clustered patterns graph G is to the left of F while for regular patterns F is to the left from G. These are distance characteristics while pcf is a second-order characteristic, being identically equal to 1 under CI. Peaks of pcf correspond to typical distances between pairs of points.

The edge-corrected estimators of these functions (Kaplan-Meier estimators for F, G, Ripley's estimator for pcf, using SPATSTAT package in R language) are obtained. We do the estimation in each of n windows for both microstructures A, B. From the estimated curves we construct vectors $G_j = G(j\triangle)$, $F_j = F(j\triangle)$, $pcf_j = pcf(j\triangle)$, $j = 1,\ldots,k$, where $\triangle > 0$ is a given step and $k\triangle$ the range considered. Choosing m of these three characteristics we construct $2n$ vectors of size km, cf. (2). If the number of independent windows n is smaller than 120, we evaluate $\mathfrak{N}$ in (2). Using K random permutations of these $2n$ vectors we perform the permutation test of

H_0 : microstructures A, B have the same distribution of characteristics involved,

exactly as in the method (I) above. If $n > 120$, we evaluate (5) and use the Kolmogorov-Smirnov test. The tests are not much dependent on $\triangle$ if it is small. All this is called method (II).

The characteristics of the spatial distribution are dependent of the intensity λ_A, λ_B of particle centroids of microstructures A, B, respectively. The intensity is the mean number of particle centroids per unit volume and it is estimated as

$$\lambda_A = \frac{n_A}{|W|}, \quad \lambda_B = \frac{n_B}{|W|},$$

where $|W|$ is the size of the window and n_A, n_B are the corresponding numbers of particle centroids observed in the window. Clearly, microstructures with different intensities have different nearest neighbour distances, etc. Consider the problem of an investigation of the difference between A and B purely in the spatial distribution independently of the different intensities. For the case $n_A \neq n_B$, we can scale the image window B by $\sqrt{n_B/n_A}$. This transformation leads to the same estimated density of particles of A and transformed B (in windows of different size). Then we evaluate functions F, G, pcf and continue by testing based on $\mathfrak{N}$-distances as above, this is our method (III).

Consider finally the problem of testing the difference in the density of particles. Here we need a parametric model and we restrict it to a point process model with no interactions (Poisson process). Under the assumption that both microstructures can be modeled by a stationary Poisson process, there is a theoretical test of the hypothesis: $H_0 : \lambda_A = \lambda_B$. We reject H_0 at a confidence level α, if the statistics:

$$T = \frac{|n_A - n_B|}{\sqrt{n_A + n_B}} > u_{1-\frac{\alpha}{2}} \tag{5}$$

(see (Ng et al, 2007)), where for $0 < a < 1$, u_a denotes the a-quantile of the standard Gaussian distribution.

2.3 Functional data approach

In Subsection 2.2 we dealt in fact with functions (F, G, pcf) which describe the spatial distribution of particles. Since in the computer we have always discrete data, i.e. a finite number of the values of a function, typically at equidistant argument points, we used the vector analysis for testing the null hypothesis by means of $\mathfrak{N}$-distances. However, within this theory it is also possible to deal with functions, this approach belongs to the field of statistical analysis of functional data.

In the functional data approach we used the Bernstein polynomials (Korovkin, 2001) as a suitable approximation for corresponding functions F, G, pcf. Let us remind that the Bernstein polynomial of the degree n for the function $f(x)$ is defined as

$$B_n(x;f) = \sum_{j=0}^{n} f\left(\frac{j}{n}\right) \binom{n}{j} x^j (1-x)^{n-j}.$$

The test for the null hypothesis is constructed by means of $\mathfrak{N}$-distances, with the strongly negative definite kernel $\mathcal{L}$ for two functions $f(x)$ and $g(x)$ defined on an interval $[0, a]$ as

$$\mathcal{L}(f, g) = \left(\int_0^a (f(x) - g(x))^2 dx \right)^{1/2}.$$

In this case the empirical analog on $\mathfrak{N}$-distance is defined as

$$\mathfrak{N} = \left(\frac{2}{M^2} \sum_{i=1}^{M} \sum_{j=1}^{M} \mathcal{L}(F_i^A, F_j^B) - \frac{1}{M^2} \sum_{i=1}^{M} \sum_{j=1}^{M} \mathcal{L}(F_i^A, F_j^A) - \frac{1}{M^2} \sum_{i=1}^{M} \sum_{j=1}^{M} \mathcal{L}(F_i^B, F_j^B) \right)^{1/2}. \quad (6)$$

Here F_j^A and F_j^B are the Bernstein polynomials for the function F (correspondingly, G or pcf), constructed for the jth window of microstructures A and B.
To compare the microstructure A with microstructure B we use permutation test, that is we combine the functions F_j^A and F_j^B in one long vector of functions, make a random permutation, and after that we split the vector into two parts, calculating after that $\mathfrak{N}$-distance between corresponding parts. The described operation has to be repeated many times, which is possible thanks to fast computers. This ends our method (IV).
Further, using the functional data approach, we suggest a new comparison technique, qualitatively different from the previous ones. Suppose again we have two microstructures (A and B), observed in M windows. Denote by n_j the number of particles in j-th window from microstructure A, and by k_j the corresponding number from B. Corresponding coordinates of particle centroids are denoted by $(X_1^{(j)}, Y_1^{(j)}), \dots, (X_{n_j}^{(j)}, Y_{n_j}^{(j)}))$ for jth window from A, and by $(U_1^{(j)}, V_1^{(j)}), \dots, (U_{k_j}^{(j)}, V_{k_j}^{(j)}))$ for jth window from B $(j = 1, 2, \dots, M)$.
Method (V) is based on a smoothing procedure in each window by convolving a discrete two-dimensional distribution concentrated in particle centroids with two-dimensional Gaussian distribution with zero mean vector and standard deviations $\sigma_j(A) = 1/\sqrt[4]{n_j}$ and $\sigma_j(B) = 1/\sqrt[4]{k_j}$, i.e. we pass to the functions

$$\mu_j(x, y) = \frac{1}{\sigma_j^2(A)} \frac{1}{n_j} \sum_{s=1}^{n_j} K\left(\frac{x - X_s^{(j)}}{\sigma_j(A)}, \frac{y - Y_s^{(j)}}{\sigma_j(A)} \right) \quad (7)$$

for A, and

$$v_j(x, y) = \frac{1}{\sigma_j^2(B)} \frac{1}{k_j} \sum_{s=1}^{k_j} K\left(\frac{x - U_s^{(j)}}{\sigma_j(B)}, \frac{y - V_s^{(j)}}{\sigma_j(B)} \right) \quad (8)$$

for B. Without loss of generality, we may suppose that the window is a unit square: $Q = \{(x, y) : 0 < x < 1, 0 < y < 1\}$. Define strongly negative definite kernel $\mathcal{L}$ for two functions $f(x, y)$ and $g(x, y)$ given on Q as

$$\mathcal{L}(f, g) = \left(\int_0^1 \int_0^1 (f(x, y) - g(x, y))^2 dx dy \right)^{1/2}.$$

In this case the empirical analog on $\mathfrak{N}$-distance (6) is used with μ_i, v_j as arguments of $\mathcal{L}$.
To compare the microstructures A and B we use again the permutation test, that is we combine the functions μ_j and v_j in one long vector, make a random permutation, and after that we split the vector into two parts, calculating after that $\mathfrak{N}$-distance between corresponding parts. The described operation has to be repeated many times.

It is possible to study a more general situation. Namely, we need not consider only coordinates (X, Y) of the particle centroids, but also individual characteristics of the particles. Then, we will have instead of two-dimensional vector (X, Y) the vectors of higher dimensionality, and instead of the functions μ and ν depending on two arguments we will have corresponding functions depending on three or more arguments. Theoretically, it is possible to consider an arbitrary number of characteristics of the particles, but the calculations for more than three arguments are very time consuming. Therefore, as the method (VI), we consider the case of three arguments, and as an example the case of three parameters: two coordinates of the particle centroid and the area of the particle (section).

Again, we have two microstructures (A and B), observed in M windows. Denote by n_j the number of particles in j-th window from microstructure A, and by k_j the corresponding number from the microstructure B. Corresponding coordinates of particle centroids and their areas are now denoted by $(X_1^{(j)}, Y_1^{(j)}, Z_1^{(j)}), \ldots, (X_{n_j}^{(j)}, Y_{n_j}^{(j)}, Z_{n_j}^{(j)})$ for j-th window from A, and by $(U_1^{(j)}, V_1^{(j)}, W_1^{j}), \ldots, (U_{k_j}^{(j)}, V_{k_j}^{(j)}, W_{k_j}^{(j)})$ for j-th window from B ($j = 1, 2, \ldots, M$).

We make a smoothing procedure in each window by convolving a discrete three-dimensional distribution concentrated in particle centroids and their areas with three-dimensional Gaussian distribution with zero mean vector and standard deviations $\sigma_j(A) = 1/\sqrt[4]{n_j}$ and $\sigma_j(B) = 1/\sqrt[4]{k_j}$, i.e. we pass to the functions

$$\mu_j(x, y, z) = \frac{1}{\sigma_j^3(A)} \frac{1}{n_j} \sum_{s=1}^{n_j} K\left(\frac{x - X_s^{(j)}}{\sigma_j(A)}, \frac{y - Y_s^{(j)}}{\sigma_j(A)}, \frac{z - Z_s^{(j)}}{\sigma_j(A)}\right)$$

for A, and

$$\nu_j(x, y, z) = \frac{1}{\sigma_j^3(B)} \frac{1}{k_j} \sum_{s=1}^{k_j} K\left(\frac{x - U_s^{(j)}}{\sigma_j(B)}, \frac{y - V_s^{(j)}}{\sigma_j(B)}, \frac{z - W_s^{(j)}}{\sigma_j(B)}\right)$$

for B.

Without loss of generality, we may suppose that the window is a unit cube: $Q = \{(x, y) : 0 < x < 1, 0 < y < 1, 0 < z < 1\}$.

Define strongly negative definite kernel $\mathcal{L}$ for two functions $f(x, y, z)$ and $g(x, y, z)$ given on Q as

$$\mathcal{L}(f, g) = \left(\int_0^1 \int_0^1 \int_0^1 (f(x, y, z) - g(x, y, z))^2 dxdydz\right)^{1/2}.$$

Then again the empirical analog on $\mathfrak{N}$-distance (6) is used with μ_i, ν_j as arguments of $\mathcal{L}$.

Further, we apply the same testing procedure as for two-dimensional case, described above.

2.4 The power of the test

In order to understand and describe the properties of the suggested testing based on $\mathfrak{N}$-distances, it is necessary to study the power of the tests, which quantifies the probability of a correct rejection of H_0. It is possible to do this by means of simulations of special cases. For the use of Kolmogorov-Smirnov test for comparison of U, V in (12) a study of the power is presented in (Klebanov, 2005). We present here another study of the power of our test when using permutation testing in the vector approach.

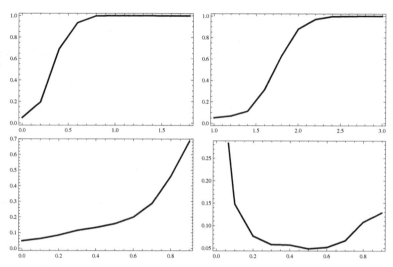

Fig. 2. Estimated probability, that the null hypothesis is rejected by the permutation test on 0.05 significance level, given the alternative of location (upper left), scale (upper right), correlation (lower left and right).

Consider a k–dimensional Gaussian distribution with mean $(\mu, \ldots, \mu)$, and variance matrix terms $\Sigma_{ii} = \sigma^2$, $\Sigma_{ij} = \rho\sigma^2$, $i \neq j$, $\rho \in [0, 1]$. This distribution can be simulated as

$$X_j = \sigma\sqrt{\rho}Z + \sigma\sqrt{1 - \rho}Y_j + \mu, \; j = 1, \ldots, k,$$

where $Z, Y_1, \ldots, Y_k$ are independent identically distributed standard Gaussian random variables. Two independent random samples of size n, with parameters $(\mu_1, \sigma_1, \rho_1), (\mu_2, \sigma_2, \rho_2)$, respectively are compared with null hypothesis of equal distributions and alternatives of
(i) location: $\mu_1 \neq \mu_2$.   (ii) scale: $\sigma_1 \neq \sigma_2$.   (iii) correlation: $\rho_1 \neq \rho_2$.
Numerical results of the simulation and testing are in Fig. 2, where it is $\mu_1 = 0$, μ_2 horizontal axis (upper left); $\sigma_1 = 1$, σ_2 horizontal axis (upper right); $\rho_1 = 0$, ρ_2 horizontal axis (lower left); $\rho_1 = 0.5$, ρ_2 horizontal axis (lower right). The number of windows is n =20, k =100 grains, 100 permutations, averaged over 1000 simulations. In the lower right graph the number of windows is 40. The parameters not involved in the alternative are $\mu_1 = \mu_2 = 0$ in (ii), (iii), $\sigma_1 = \sigma_2 = 1$ in (i), (iii), $\rho_1 = \rho_2 = 0$ in (i), (ii). We can observe that for the location alternative the power function increases more rapidly than for the scale and correlation alternatives. For both location and scale alternatives the power function increases more rapidly than for the correlation alternatives.

3. Materials

Further we present an application of suggested statistical methods. A Czech company AL INVEST Břidličná, a.s. provided five Al-Mn alloys denoted A, C, L, P, Z, the composition of which is in Table 1. The alloys with high manganese contents are A, C, L, P, high silicon contents have A, C and L, they differ in the zinc contents present in C and not present in A, L. Considering the high solubility of Zn in Al, all Zn is dissolved in aluminium matrix. Therefore,

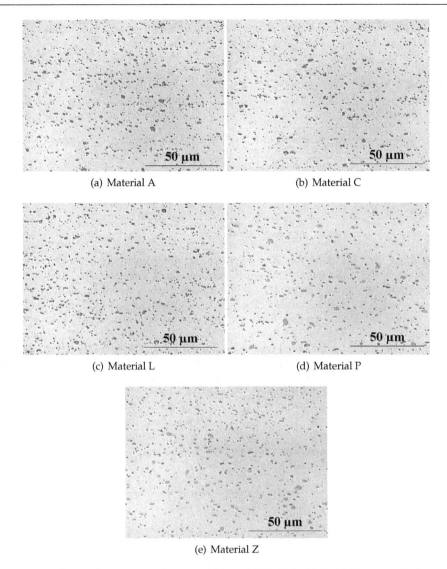

(a) Material A   (b) Material C

(c) Material L   (d) Material P

(e) Material Z

Fig. 3. A metallographic sample of material (a) A, (b) C, (c) L, (d) P, (e) Z, respectively, a transverse section of the foil. In (a), (b), (c), (alloys with higher contents of Si) particles α-Al_{12}(Mn, Fe)$_3$Si prevail. In (d), (e), (alloys with lower contents of Si) particles Al_6(Mn, Fe) prevail.

Zn does not participate in second-phase particles and its presence in alloy C does not affect its particle volume fraction and size distribution. Alloys P and Z have lower silicon contents, P has lower copper contents while Z has lower manganese contents. The most important factor influencing particles volume fraction and size distribution in the set of alloys considered is the content of silicon. Coarse particles are mostly primary, undissolved particles of α-Al_{12}(Mn,

Fe)$_3$Si and Al$_6$(Mn, Fe). Fine particles, present especially for alloys with a higher content of Si, are mainly precipitates α-Al$_{12}$(Mn, Fe)$_3$Si, cf. (Slámová et al., 2006).

All alloys were twin-roll cast in strips of 8.5 mm in thickness. All specimens were homogenized at high temperature after a 35% reduction in thickness and then cold rolled to thickness of 0.4 mm. The samples of 0.4 mm thickness were annealed again at 350 °C in order to increase the ductility of the material so as to facilitate the cold rolling up to the final foil thickness of 0.10 mm.

Alloy	Mn	Si	Fe	Cu	Zn	Mn+Si+Fe	(Mn+Si)/Fe	Mn/Si
A	1.09	0.54	0.47	0.16	0.004	2.1	3.5	2.02
C	1.02	0.56	0.53	0.15	1.02	2.1	3.1	1.8
L	1.02	0.59	0.48	0.11	0.01	2.1	3.3	1.7
P	1.01	0.20	0.61	0.04	-	1.8	2.0	5.1
Z	0.86	0.10	0.61	0.16	-	1.6	1.6	8.7

Table 1. Composition of experimental materials [wt. %].

Metallographic samples after grinding and polishing were observed by optical microscopy, see Fig. 3, where the vertical size of the window corresponds to the total thickness of the foil. Twenty windows along the foil, see Fig. 4, with distance 1 mm between neighbouring ones were measured by an image analyzer. The number of particles and individual parameters of each particle were measured in each window and for each material. The individual parameters are Area, EqDiam (i.e. diameter of the circle with the same area as parameter Area), Minimal Feret, Maximal Feret (extremes of the breadth, Ohser & Mücklich (2000), of a particle w.r.t. directions). The Shape Factor can be evaluated as a fraction of Minimal Feret and Maximal Feret. Parameters Area, EqDiam and Shape Factor were used to the input data to our test.

Fig. 4. Scheme of sampling windows along the foil of thickness 0.1 mm. The distance between neighbouring windows is 1 mm.

4. Numerical results

The preliminary analysis concerns the mean size of particle sections and their density, see Tables 2, 3, apparently A, C, L differ from P and Z, see also Fig.5. For the testing, since the number of windows $n = 20$ is small, we use the permutation test described in Section 2. The results of method (I) are in Table 4. The microstructures do not differ in the shape factor of particles. Concerning the particles size we observe again two separated groups $\{A, C, L\}$ and $\{P, Z\}$ (according to silicon contents). Between them there is a significant difference, while within the groups this is not the case. Nevertheless in Table 4 we observe a difference in pairs $A - L$, $P - Z$, while we cannot reject the null hypothesis in pairs $A - C$, $C - L$.

For the methods (II)-(IV) of the spatial distribution of particle section centroids first the F, G and pcf functions were estimated. The estimators of functions F, G and pcf for all windows of material P are presented in Fig. 6, i.e. in each figure a), b), c) there are 20 graphs. We observe a small variability of the estimators. Let us note that there is a similarly small variability of these estimators in all other materials. In Fig. 7 we compare the average estimators (from

Alloy	MeanArea	Standard Deviation	Density	Total number
A	0.696	0.864	0.073	19 773
C	0.683	0.873	0.070	18 834
L	0.689	0.796	0.080	21 611
P	1.080	1.287	0.047	12 605
Z	1.223	1.386	0.041	10 997

Table 2. The mean area of observed particle sections (in $[\mu m^2]$), their density (in $[\mu m^{-2}]$) and total number for materials A, C, L, P, Z, evaluated from all 20 windows.

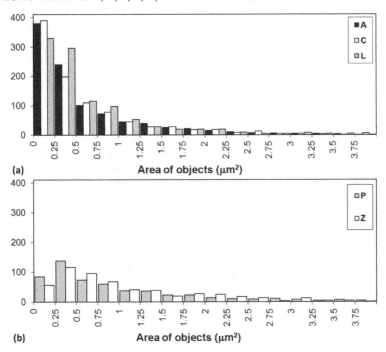

(a)   Area of objects (μm^2)

(b)   Area of objects (μm^2)

Fig. 5. Histograms of areas of particle sections in the first window of each material. In a) we have A, C, L and in b) there are P and Z.

Alloy	Mean Number	Standard Deviation
A	988,65	44.76
C	941,7	53.50
L	1080,55	41.36
P	630,25	46.55
Z	549,85	33.47

Table 3. The mean number of observed particle sections and their standard deviation in one window (evaluated from all 20 windows).

20 windows) of the functions F, G and pcf evaluated for all materials. We obtain the results that for the material L, A, C these are greatly different from the results for materials P and Z. The estimators of pcf of materials P and Z are practically the same and the differences of

	All	Area	Eqdiam	ShapeFactor
A-C	0.343	0.369	0.335	0.050
A-L	0.040	0.042	0.035	0.203
A-P	< 0.001	< 0.001	< 0.001	0.565
A-Z	< 0.001	< 0.001	< 0.001	0.529
C-L	0.302	0.355	0.288	0.046
C-P	< 0.001	< 0.001	< 0.001	0.283
C-Z	< 0.001	< 0.001	< 0.001	0.322
L-P	< 0.001	< 0.001	< 0.001	0.422
L-Z	< 0.001	< 0.001	< 0.001	0.488
P-Z	0.015	0.020	0.002	0.639

Table 4. The p-values for two-sample tests of individual particle parameters (method I) evaluated for all $m = 3$ parameters together (the column All) and then for single parameters ($m = 1$) with $n = 20, k = 500$ and with the number of permutations equal to 5000.

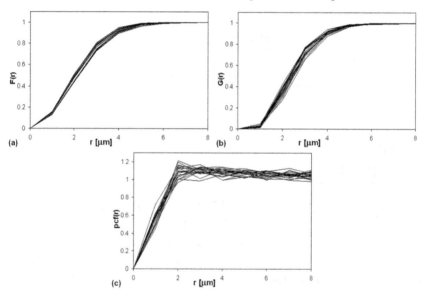

Fig. 6. Estimators of (a) F-function, (b) G-function and (c) pcf for material P. Graphs obtained from each of 20 windows are drawn in the same figures in order to observe a small variability of the estimators among the windows.

estimators of the functions F and G are small, while the corresponding functions of materials L, A, C are shifted to the left. This is caused mostly by the different particle density
The results for method (II) of the two-sample tests for the spatial distribution in Table 5 lead to the interpretation that there are significant differences in spatial distribution between any materials of different groups $\{A, C, L\}$ and $\{P, Z\}$. It is interesting to observe what happens within the groups. We can see that while distribution functions F, G still yield differences, the pair correlation function does not reveal any. Only the pair $C - A$ which has the most close contents of silicon, does not reveal any difference in any characteristics.

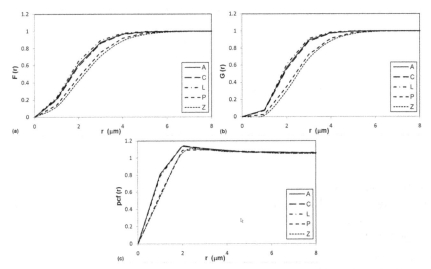

Fig. 7. Graphs of (a) F-function, (b) G-function and (c) pcf for all three materials obtained by averaging the estimators from 20 windows. We observe that for materials L, A, C they look differently from P, Z.

	All	F	G	pcf
A-C	0.6	0.166	0.034	0.999
A-L	0.004	< 0.001	< 0.001	0.784
A-P,A-Z	< 0.001	< 0.001	< 0.001	< 0.001
C-L	< 0.001	< 0.001	< 0.001	0.584
C-P,C-Z	< 0.001	< 0.001	< 0.001	< 0.001
L-P,L-Z	< 0.001	< 0.001	< 0.001	< 0.001
P-Z	< 0.001	< 0.001	< 0.001	0.545

Table 5. The p-values for two-sample tests of the spatial distribution (method II) evaluated for all three parameters together ($m = 3$, column All) and then for single parameters ($m = 1$) with $n = 20$, $k = 11$, $\triangle = 1\ \mu m$ and with the number of permutations equal to 5000.

Let us analyze the number of particles for the microstructures P and Z, we can test the difference between $n_P = 13866$, $n_Z = 11945$ from all 20 windows, thus in (5) we have $T = 12 > 1.96$ and we reject an H_0 of equal particle density at the significance level $\alpha = 0.05$. Here the Poisson process model assumption is violated, as suggested by the shape of pcf in Fig. 6 we have a type of a regular model, i.e. mild repulsion since there are nonoverlapping particles around the centroids. Clearly, if we reject the null hypothesis $\lambda_P = \lambda_Z$ for the Poisson process model using (5), we reject it for the regular model too, since it is less dispersed, i.e. the numbers of particles observed in windows vary more slowly.

Further a finer analysis of the spatial distribution of particles is applied using method (III). If we eliminate the effect of particle density on the spatial distribution by means of the scale change as suggested in Section 2, the results change as presented in Table 6. The pure effect of the spatial distribution of particle centroids is such that there is no significant difference between materials $P - Z$, $A - C$. But this is moreover the case also for individual functions F

	All	F	G	pcf
A-C	0.483	0.132	0.527	0.540
A-L	0.003	0.002	0.509	0.004
A-P	0.019	0.108	0.022	0.028
A-Z	0.013	0.178	0.122	0.009
C-L	< 0.001	< 0.001	0.502	< 0.001
C-P	0.001	0.060	0.057	0.001
C-Z	0.005	0.195	0.158	0.003
L-P	< 0.001	0.010	0.037	0.001
L-Z	< 0.001	0.005	0.256	< 0.001
P-Z	0.628	0.611	0.416	0.479

Table 6. The p-values for two-sample tests of the spatial distribution (with the effect of particle density eliminated, method III). Evaluation for all three parameters together (first column) and then for single parameters ($m = 1$) was obtained with $n = 20$, $k = 11$, $\triangle = 1\ \mu m$ and with the number of permutations equal to 5000.

(holds for $A - P$, $A - Z$, $C - Z$) and G ($A - L$, $A - Z$, $C - L$, $C - Z$, $L - Z$). That means some differences between two groups are removed.

Finally we present results of two-sample tests when using the functional data approach in Subsection 2.3. First for the comparison based on functions F, G, pcf we use the versions with scale change to eliminate the effect of particle density. Similar results as in Table 6 are expected using this method (IV), see Table 7. Even if individual p-values in both tables differ, the decisions about H_0 are almost completely the same.

	F	G	pcf
A–C	0.59	0.57	0.32
A–L	0.02	0.46	< 0.01
A–P	0.17	0.02	0.02
A–Z	0.42	0.13	< 0.01
C–L	< 0.01	0.42	< 0.01
C–P	0.07	0.04	< 0.01
C–Z	0.43	0.14	0.01
L–P	0.02	0.04	< 0.01
L–Z	0.01	0.26	< 0.01
P–Z	0.58	0.44	0.46

Table 7. The p-values for two-sample tests of the spatial distribution (with the effect of particle density eliminated), using the functional data approach, method (IV). Evaluation was obtained with the number of permutations equal to 100.

Further the tests based on multidimensional smoothing of particle characteristics are applied, which are qualitatively different methods. They are not sensitive to the particle density, on the other hand it may reveal local inhomogeneities and differences. First we give the results of comparison by method (V), that is corresponding p-values of the test based on $\mathfrak{N}$-distances of functions (7), (8), for different pairs of microstructures in terms of the particle centroid coordinates only, see Table 8. In many cases, but not in all, the results are similar to those in Tables 6, 7 (spatial distribution only is investigated in both methods). Different results are obtained especially for pairs $A - C$, $P - L$.

A-C	A-L	A-P	A-Z	C-L	C-P	C-Z	L-P	L-Z	P-Z
< 0.01	0.0297	< 0.01	0.475	< 0.01	< 0.01	< 0.01	0.257	< 0.01	0.059
< 0.01	0.07	< 0.01	0.19	< 0.01	< 0.01	< 0.01	0.11	0.02	0.04

Table 8. The p-values of $\mathfrak{N}$-test for corresponding coordinates comparisons in the functional data approach, method (V). The top row corresponds to the choice $\sigma_j(A) = 1/\sqrt[4]{n_j}$ and $\sigma_j(B) = 1/\sqrt[4]{k_j}$, the bottom row to the choice $\sigma_j(A) = \sigma_j(B) = \frac{1}{3}$.

A-L	A-P	P-Z
< 0.01	< 0.01	< 0.01

Table 9. The p-values of $\mathfrak{N}$-test for the comparisons based on coordinates and areas of the particles in the functional data approach, method (VI).

Finally a simultaneous analysis of spatial distribution and an individual particle parameter (area of the section) was performed using the functional data approach, method (VI). The results of the comparison are given in Table 9. We do not consider all pairs of microstructures since for those of different groups (different silicon contents) the particle areas surely cause the rejection of null-hypothesis. As we can see from Table 9, this is the case also in other pairs.

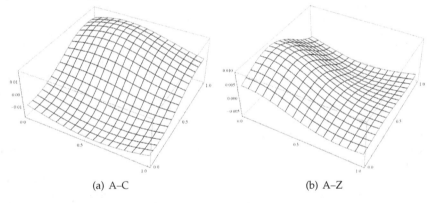

(a) A–C   (b) A–Z

Fig. 8. Average value of the difference $\mu_j - \nu_j$ of functions (7),(8) taken from all 20 windows, microstructure A–C, A–Z.

5. Concluding remarks

This chapter brings an extension and continuation of research started in (Benes et al., 2009). New statistical methods are developed for the comparison of microstructural images of random objects in metallography and other applications. They are based on an appropriate interaction of approaches from mathematical statistics, image analysis and stochastic geometry. A proper two-sample test derived from $\mathfrak{N}$-distances enables one to evaluate a large amount of information from a few observed windows. The tests presented are easy to apply to metallographic images observed by light microscopy and image analysis. In comparison with the above mentioned paper here we suggest further methods based on functional data analysis and we analyze a broader set of foils from aluminium-manganese based alloys.

The first group of methods based on vectors of characteristics obtained from image analysis measurements and further transformation of data is well-established and the results easily understandable. From the practical point-of-view, it should be mentioned that $\mathfrak{N}$-distances are scale dependent, so that when evaluating qualitatively different information simultaneously ($m > 1$) one has to choose the scale carefully to be comparable for all parameters. This is guaranteed in the analysis of spatial distribution since all three functions used fall within a similar range. In the analysis of particle characteristics the size scale has to be modified comparably to the range of the shape factor. We conclude that mostly recommended methods are (I) and (IV) (or (III)).

On the other hand the newly proposed methods (V) and (VI) based on multidimensional functional data analysis need further investigations to be able to claim their usefulness. In comparison with method (I) they are able to evaluate different numbers of particles in windows. Theoretically, they are independent on the dimensionality, and therefore they are of great potential use in multidimensional statistical analysis. At first sight corresponding tests seem to be strict and sensitive to both local differences in spatial distribution and in particle characteristics. One must be very careful when interpreting the results of a functional data analysis. It should be also added that functional data analysis combined with permutation testing is more time-consuming (especially in the multivariate case), but feasible when using fast computers, and, especially, clusters.

Some of results obtained by classical and functional data analysis may seem to be contradictory, but this is not the case, the reason is that various methods for comparison of spatial distribution are different in their nature. We may consider pairs of microstructures A-C and A-Z. The functional data method using functions μ, ν in (7),(8) rejects the null hypothesis for A-C and does not reject it for A-Z. This conclusion is related to Fig. 8, where we can see the average values of the difference $\mu_j - \nu_j$ taken from all 20 windows, where the range is two times smaller for A-Z than for A-C. The same observation holds for individual windows, too. We investigated also the sensitivity of method (V) with respect to the choice of bandwidth σ_j. In the top row of $p-$values in Table 8 there is the asymptotically optimal bandwidth by theory. One can observe how the $p-$value slightly changes with a broader bandwidth in bottom row, but there is no evidence of a systematic change.

It is possible to study statistical properties of the tests by simulations. Concerning the resolution of the test, the power of the variants of the test based on $\mathfrak{N}$-distances was compared in (Klebanov, 2005). Besides our study in Subsection 2.4, in paper (Bakshaev, 2008) a large comparative study of the power of several two-sample tests (Kolmogorov-Smirnov, Cramer-von Mises, Anderson-Darling, Wilcoxon, Mann-Whitney, $\mathfrak{N}$-distances) was made. It appears that for a multidimensional case the test based on $\mathfrak{N}$-distances has the highest power. The results of the application of the two-sample test in metallography can be transformed from conclusions about the geometry of the microstructure to conclusions relevant materials research. Since the production of all three materials was based on the same processing, the only difference is in the chemical composition of the alloys. Therefore, we can conclude that a differentiation for high and low silicon contents is apparent, while small differences in composition within different groups $\{A, C, L\}$ and $\{P, Z\}$ do not have a clearly apparent impact on the microstructure. We can observe that while P and Z have different particle densities, they do not differ in particle size and shape, nor in the pure spatial distribution (interactions).

6. Acknowledgement

The research was supported by the Czech Science Foundation, project GAČR P201/10/0472, and by the Czech Ministery of Education, project MSM 0021620839. Our memory belongs to Margarita Slámová, who was the coauthor of the paper (Benes et al., 2009). She died prematurely in 2009 and we missed her during the preparation of this chapter a lot.

7. Appendix

Here we give a mathematical background of $\mathfrak{N}$-distances and related statistical testing. This background comes from (Klebanov, 2005). Let $\{\mathfrak{X}, \mathfrak{A}\}$ be a measurable space, $\mathcal{L} : \mathfrak{X}^2 \to \mathbb{R}^1$ is a negative definite kernel on $\mathfrak{X}$ ($\mathcal{L}(x,x) = 0$ and $\mathcal{L}(x,y) = \mathcal{L}(y,x)$) if and only if

$$\int_{\mathfrak{X}} \int_{\mathfrak{X}} \mathcal{L}(x,y)h(x)h(y)dQ(x)dQ(y) \leq 0 \qquad (9)$$

for an arbitrary probability measure Q on $\{\mathfrak{X}, \mathfrak{A}\}$ and a measurable function h on $\mathfrak{X}$ such that $\int_{\mathfrak{X}} h(x)dQ(x) = 0$. We say that $\mathcal{L}$ is strongly negative definite if the equality in (9) implies $h = 0$ almost surely with respect to the measure Q.

Let $\mathcal{L}$ be a strongly negative definite kernel on $\mathfrak{X}$, $\mathcal{B}_{\mathcal{L}}$ the set of all probabilities μ on $\{\mathfrak{X}, \mathfrak{A}\}$ for which there exists the integral $\int_{\mathfrak{X}} \int_{\mathfrak{X}} \mathcal{L}(x,y)d\mu(x)d\mu(y) < \infty$. For $\mu, \nu \in \mathcal{B}_{\mathcal{L}}$ put $\mathcal{N}(\mu,\nu) =$

$$= 2\int_{\mathfrak{X}} \int_{\mathfrak{X}} \mathcal{L}(x,y)d\mu(x)d\nu(y) - \int_{\mathfrak{X}} \int_{\mathfrak{X}} \mathcal{L}(x,y)d\mu(x)d\mu(y) - \int_{\mathfrak{X}} \int_{\mathfrak{X}} \mathcal{L}(x,y)d\nu(x)d\nu(y). \qquad (10)$$

Then

$$\mathfrak{N}(\mu,\nu) = \left(\mathcal{N}(\mu,\nu)\right)^{1/2} \qquad (11)$$

is a distance on $\mathcal{B}_{\mathcal{L}}$, it is called $\mathfrak{N}$-distance. We will use $\mathfrak{N}$-distances in the following using two approaches. First in classical data analysis when $\mathfrak{X} = \mathbb{R}^k$ is the Euclidean space of k-dimensional vectors. Secondly in functional data analysis where $\mathfrak{X} = L_2$ is the space of square integrable functions.

Let $\mathcal{L}(x,y)$ be a strongly negative definite kernel on $\mathbb{R}^k$, X, Y are two independent random vectors in $\mathbb{R}^k$, define one-dimensional independent random variables U, V by

$$U = \mathcal{L}(X,Y) - \mathcal{L}(X,X'), \qquad V = \mathcal{L}(Y',Y'') - \mathcal{L}(X'',Y''). \qquad (12)$$

Here all vectors X, X', X'', Y, Y', Y'' are mutually independent, equalities of distributions $X \overset{d}{=} X' \overset{d}{=} X''$, $Y \overset{d}{=} Y' \overset{d}{=} Y''$ hold. We have $X \overset{d}{=} Y \iff U \overset{d}{=} V \iff \mathfrak{N}(X,Y) = 0$. Consider testing of the hypothesis $H_0 : X \overset{d}{=} Y$ for multivariate random vectors X, Y. This hypothesis is equivalent to $H_0' : U \overset{d}{=} V$, where U, V are random variables taking values in $\mathbb{R}^1$. Consider two independent samples $X_1, \ldots, X_n$; $Y_1, \ldots, Y_n$ from general multivariate populations X and Y, respectively. A one-dimensional test to U and V can proceed in the following ways:
a) split each sample randomly in three equal parts X, X', X'', Y, Y', Y'' and use (12); this leads to a loss of information,
b) simulate the samples from X' and X'' (as well as from Y' and Y'') by independent choices from observations $X_1, \ldots, X_n$ (and from $Y_1, \ldots, Y_n$, correspondingly); thus we do not test the hypothesis $X \overset{d}{=} Y$, but the one of the corresponding empirical distributions,
c) permutation test using Monte Carlo approximation.

8. References

Bakshaev A. (2008). Nonparametric tests based on N-distances, *Lithuanian Math J*, Vol. 48, No. 4, 357–376.

Beneš V, Lechnerová R, Klebanov L, Slámová M, Sláma P. (2009). Statistical comparison of the geometry of second-phase particles. *Mater Charact*, Vol. 60, 1076-1081.

Buening H, Trenkler G. (1978). *Nonparametric statistical methods*, Walter de Gruyter, Berlin (in German).

Derr R, Ji C. (2000). *Fitting microstructural models in materials science.* http://www.stat-or.unc.edu/webspace/miscellaneous/cji/scan05.pdf.

Hirsch J. (2006). *Virtual fabrication of aluminium products*, John Wiley & Sons, New York.

Humphreys FJ, Hatherly M. (2004). *Recrystallization and related annealing phenomena*, Elsevier, London.

Ihaka R, Gentleman R. (1996). R: A language for data analysis and graphics, *J Comput Graph Stat*, Vol. 5, No. 3, 299–314.

Klebanov LB. (2005). $\mathfrak{N}$-distances and their applications, The Karolinum Press, Prague.

Korovkin PP. (2001). *Bernstein Polynomials*, In: Hazewinkel M, Encyclopaedia of Mathematics, Springer, New York.

Kupczyk J. (2006). Application of significance tests in quantitative metallographic analysis of a C-Mn-B steel. *Mater Charact*, Vol. 57, 171–175.

Lehmann E, Romano P. (2005). *Testing statistical hypotheses*, Springer, New York.

Ng HKT, Gu K, Tang ML (2007). A comparative study of tests for the difference of two Poisson means. *Comp Stat & Data Anal*, Vol. 51, 3085–3099.

Ohser J, Mücklich F. (2000). *Statistical analysis of microstructures in materials science*, Wiley.

Polmear IJ. (2006). *Light alloys - from traditional alloys to nanocrystals*, The Fourth edition, Elsevier, London.

Slámová M, Sláma P, Cieslar M. (2006). *The influence of alloy composition on phase transformations and recrystallization in twin-roll cast AlMnFe alloys*, In: Poole WJ, Wells MA, Lloyd DJ, editors. Aluminium alloys. Mat. Sci. Forum, Vol. 519-521, 365–370.

Tewari A, Gokhale AM. (2006). Computations of contact distributions for representation of microstructural spatial clustering. *Comput Mater Sci*, Vol. 38, 75–82.

Tewari A, Gokhale AM. (2006). Nearest neighbor distributions in thin films, sheets, and plates. *Acta Mater*, Vol. 54, No. 7, 1957–63.

Permissions

The contributors of this book come from diverse backgrounds, making this book a truly international effort. This book will bring forth new frontiers with its revolutionizing research information and detailed analysis of the nascent developments around the world.

We would like to thank Dr. Zaki Ahmad (Professor Emeritus), for lending his expertise to make the book truly unique. He has played a crucial role in the development of this book. Without his invaluable contribution this book wouldn't have been possible. He has made vital efforts to compile up to date information on the varied aspects of this subject to make this book a valuable addition to the collection of many professionals and students.

This book was conceptualized with the vision of imparting up-to-date information and advanced data in this field. To ensure the same, a matchless editorial board was set up. Every individual on the board went through rigorous rounds of assessment to prove their worth. After which they invested a large part of their time researching and compiling the most relevant data for our readers. Conferences and sessions were held from time to time between the editorial board and the contributing authors to present the data in the most comprehensible form. The editorial team has worked tirelessly to provide valuable and valid information to help people across the globe.

Every chapter published in this book has been scrutinized by our experts. Their significance has been extensively debated. The topics covered herein carry significant findings which will fuel the growth of the discipline. They may even be implemented as practical applications or may be referred to as a beginning point for another development. Chapters in this book were first published by InTech; hereby published with permission under the Creative Commons Attribution License or equivalent.

The editorial board has been involved in producing this book since its inception. They have spent rigorous hours researching and exploring the diverse topics which have resulted in the successful publishing of this book. They have passed on their knowledge of decades through this book. To expedite this challenging task, the publisher supported the team at every step. A small team of assistant editors was also appointed to further simplify the editing procedure and attain best results for the readers.

Our editorial team has been hand-picked from every corner of the world. Their multi-ethnicity adds dynamic inputs to the discussions which result in innovative outcomes. These outcomes are then further discussed with the researchers and contributors who give their valuable feedback and opinion regarding the same. The feedback is then collaborated with the researches and they are edited in a comprehensive manner to aid the understanding of the subject.

Apart from the editorial board, the designing team has also invested a significant amount of their time in understanding the subject and creating the most relevant covers. They scrutinized every image to scout for the most suitable representation of the subject and create an appropriate cover for the book.

The publishing team has been involved in this book since its early stages. They were actively engaged in every process, be it collecting the data, connecting with the contributors or procuring relevant information. The team has been an ardent support to the editorial, designing and production team. Their endless efforts to recruit the best for this project, has resulted in the accomplishment of this book. They are a veteran in the field of academics and their pool of knowledge is as vast as their experience in printing. Their expertise and guidance has proved useful at every step. Their uncompromising quality standards have made this book an exceptional effort. Their encouragement from time to time has been an inspiration for everyone.

The publisher and the editorial board hope that this book will prove to be a valuable piece of knowledge for researchers, students, practitioners and scholars across the globe.

List of Contributors

Theodore E. Matikas
Department of Materials Science and Engineering, University of Ioannina, Greece

Syed T. Hasan
Faculty of Arts, Computing, Engineering and Sciences, Sheffield Hallam University, United Kingdom

Mariyam Jameelah Ghazali
Department of Mechanical & Materials Engineering, Universiti Kebangsaan Malaysia, Malaysia

Eichlseder Wilfried, Winter Gerhard, Minichmayr Robert and Riedler Martin
Montanuniversität Leoben, Austria

Zaki Ahmad and B. J. Abdul Aleem
Mechanical Engineering Department, King Fahd University of Petroleum & Minerals, Dhahran, Saudi Arabia

Amir Farzaneh
Department of Metals, International Center for Science, High Technology and Environmental Sciences, Kerman, Iran

Vladimir Kornev, Evgeniy Karpov and Alexander Demeshkin
Lavrentyev Institute of Hydrodynamics SB RAS, Russia

Ali Reza Eivani
Materials Innovation Institute (M2i), Mekelweg 2, 2628 CD Delft, The Netherlands
Department of Materials Science and Engineering, Delft University of Technology, Mekelweg 2, 2628 CD Delft, The Netherlands

Jie Zhou and Jurek Duszczyk
Department of Materials Science and Engineering, Delft University of Technology, Mekelweg 2, 2628 CD Delft, The Netherlands

Abbas Akbarzadeh
Department of Materials Science and Engineering, Sharif University of Technology, Tehran, Iran

Viktor Beneš and Lev Klebanov
Charles University in Prague, Faculty of Mathematics and Physics, Department of Probability and Mathematical Statistics, Czech Republic

Radka Lechnerová
Private College of Economic Studies, Ltd., Prague, Czech Republic

Peter Sláma
COMTES FHT a.s., Metallography, Dob˘rany, Czech Republic

Printed in the USA
CPSIA information can be obtained
at www.ICGtesting.com
JSHW011407221024
72173JS00003B/445